AF615801

Technologies for Detecting Heritable Mutations in Human Beings

Technologies for Detecting Heritable Mutations in Human Beings

OFFICE OF TECHNOLOGY ASSESSMENT TASK FORCE

Michael Gough, Project Director
Julia T. Ostrowsky, Analyst
Hellen Gelband, Analyst

Elbert Branscomb, Lawrence Livermore National Laboratory
Neal Cariello, Massachusetts Institute of Technology
Leonard Lerman, Genetics Institute
Harvey Mohrenweiser, University of Michigan Medical School
Richard M. Myers, University of California at San Francisco
Janice A. Nicklas, University of Vermont Medical School
Maynard Olson, Washington University Medical School
Cassandra Smith, Columbia University College of Physicians and Surgeons
L.H.T. Van der Ploeg, Columbia University College of Physicians and Surgeons
Diane K. Wagener, Washington, D.C.

Science Information Resource Center

J.B. LIPPINCOTT COMPANY
PHILADELPHIA

London Mexico City New York St. Louis São Paulo Sydney

Authorized Hardbound Edition, 1988

Science Information Resource Center
East Washington Square
Philadelphia, Pennsylvania 19105

Publisher's Note:

This permanent edition includes the complete text of the Office of Technology Assessment Special Report entitled *Technologies for Detecting Heritable Mutations in Human Beings*. Under the direction of the OTA Project Staff and with the help of a distinguished Advisory Panel and the OTA Health Program Advisory Committee, ten Author/Contractors prepared much new material and reviewed and distilled a wealth of existing information.

Science Information Resource Center is a cooperative venture of J.B. Lippincott Company and Hemisphere Publishing Corporation, subsidiaries of Harper & Row, Publishers, Inc., New York.

Library of Congress Cataloging-in-Publication Data
Technologies for detecting heritable mutations in human beings.
Reprint. Originally published: Washington, D.C.: Congress of the U.S., Office of Technology Assessment: For sale by the Supt. of Docs., U.S. G.P.O., [1986].
Bibliography: p.
Includes index.
1. Medical genetics—Methodology. 2. Mutation. (Biology) 3. Mutagenicity testing. I. Gough, Michael, 1939- . II. United States. Congress. Office of Technology Assessment. [DNLM: 1. Genetics, Medical. 2. Mutagenicity Testing. QZ 50 T255 1986a]
RB155.T43 1988 616'.042'028 87-26323
ISBN 0-397-53004-8

Foreword

Ensuring the health of future generations of children is of obvious importance to American society. Heritable mutations, permanent changes in the genetic material that can be passed on to succeeding generations, are the cause of a large but currently unquantifiable share of embryonic and fetal loss, disease, disability, and early death in the United States today. The methods now available to study heritable mutations, however, offer relatively little information about the kinds of mutations that can occur, their frequency, or their causes. Recent advances in molecular genetics have opened the door to new and innovative technologies that may offer a great deal more information about DNA. It may soon be possible to characterize mutations precisely, to measure their frequency, and perhaps also to associate particular mutations with exposures to specific mutagenic influences. While some of the new technologies are still on the drawing board, they are developing quickly and several of them may become available for wide-scale use in the next 5 to 10 years.

The Senate Committee on Veterans' Affairs, the House Committee on Science and Technology, and the House Committee on Energy and Commerce requested that OTA assess the available information about current and proposed means for detecting heritable mutations and on the likelihood and potential impact of such technological advances. These committees have wrestled with the problems of determining whether past exposures to potential mutagens have affected the health of Americans, and of framing reasonable public health laws, given current knowledge and technologies. This report summarizes OTA's findings as they relate to these issues.

An advisory panel, chaired by Arno G. Motulsky, provided guidance and assistance during the assessment. The OTA Health Program Advisory Committee, OTA staff, and scientific and policy experts from the private sector, academia, and the Federal Government provided information during the assessment and reviewed drafts of the report. We thank all who assisted us. As with all OTA reports, the content of the assessment is the sole responsibility of OTA and does not necessarily constitute the consensus or endorsement of the advisory panel or the Technology Assessment Board. Key OTA staff involved in the assessment were Michael Gough, Julie Ostrowsky, and Hellen Gelband.

John H. Gibbons

JOHN H. GIBBONS
Director

Advisory Panel for Technologies for Detecting Heritable Mutations in Human Beings

Arno G. Motulsky, *Panel Chair*
Center for Inherited Diseases
University of Washington School of Medicine

Richard J. Albertini
Department of Medicine
University of Vermont
College of Medicine

Michael S. Baram
School of Medicine and Public Health
Boston University

Charles R. Cantor
Department of Human Genetics and Development
Columbia University College of Physicians and Surgeons

Dale Hattis
Center for Policy Alternatives
Massachusetts Institute of Technology

Ernest B. Hook
Bureau of Maternal and Child Health
New York State Department of Health

Alfred G. Knudson, Jr.
Institute for Cancer Research
Fox Chase Cancer Center

Nan M. Laird
Department of Biological Statistics
Harvard University
School of Public Health

Mortimer L. Mendelsohn
Lawrence Livermore National Laboratory

Jeffrey H. Miller
Department of Biology
University of California at Los Angeles

James V. Neel
Department of Human Genetics
University of Michigan School of Medicine

Norton Nelson
Department of Environmental Medicine
New York University School of Medicine

Mark L. Pearson
E.I. du Pont de Nemours & Co.

Richard K. Riegelman
Department of Medicine
George Washington University School of Medicine

Liane B. Russell
Oak Ridge National Laboratory

Richard B. Setlow
Brookhaven National Laboratory

William J. Schull
The University of Texas Health Science Center

William G. Thilly
Department of Nutrition and Food Science
Massachusetts Institute of Technology

Richard M. Myers, Special Consultant
University of California at San Francisco

NOTE: OTA appreciates and is grateful for the valuable assistance and thoughtful critiques provided by the advisory panel members. The panel does not, however, necessarily approve, disapprove, or endorse this report. OTA assumes full responsibility for the report and the accuracy of its contents.

OTA Project Staff—Technologies for Detecting Heritable Mutations in Human Beings

Roger C. Herdman, *Assistant Director, OTA*
Health and Life Sciences Division

Clyde J. Behney, *Health Program Manager*

Michael Gough, *Project Director*

Julia T. Ostrowsky, *Analyst*

Hellen Gelband, *Analyst*

Other Contributing Staff

Cheryl M. Corsaro, *Analyst*[1]

Virginia Cwalina, *Administrative Assistant*

Carol Ann Guntow, *Secretary/Word Processor Specialist*

Diann G. Hohenthaner, *Word Processor/P.C. Specialist*

Eric Passaglia, *Senior Distribution Specialist*

Contractors

Elbert Branscomb, Lawrence Livermore National Laboratory

Neal Cariello, Massachusetts Institute of Technology

Leonard Lerman, Genetics Institute

Harvey Mohrenweiser, University of Michigan Medical School

Richard M. Myers, University of California at San Francisco

Janice A. Nicklas, University of Vermont School of Medicine

Maynard Olson, Washington University Medical School

Cassandra Smith, Columbia University College of Physicians and Surgeons

L.H.T. Van der Ploeg, Columbia University College of Physicians and Surgeons

Diane K. Wagener, Washington, DC

[1]On detail from the National Institutes of Health, April-June 1985.

Contents

Glossary of Acronyms and Terms

Glossary of Acronyms

A —Adenine
AEC —Atomic Energy Commission
ATSDR —Agency for Toxic Substances and Disease Registry (Centers for Disease Control)
BEIR —Committee on the Biological Effects of Ionizing Radiations (NRC)
C —Cytosine
CERCLA —Comprehensive Environmental Response, Compensation, and Liability Act ("Superfund")
DNA —Deoxyribonucleic Acid
DOE —Department of Energy
EBV —Epstein-Barr Virus
EDB —Ethylene dibromide
EPA —Environmental Protection Agency
ENU —Ethylnitrosourea
EtO —Ethylene oxide
FDA —Food and Drug Administration (DHHS)
FRC —Federal Radiation Council
G —Guanine
HbA —Hemoglobin A
HbM —Hemoglobin M
hprt —Hypoxanthineguanine phosphoribosyl transferase (gene)
HPRT —Hypoxanthineguanine phosphoribosyl transferase (enzyme)
HTT —Heritable Translocation Test
ICPEMC —International Commission for Protection Against Environmental Mutagens and Carginogens
ICRP —International Commission on Radiological Protection
mRNA —Messenger RNA
MSHA —Mine Safety and Health Administration (Department of Labor)
NCRP —National Council on Radiation Protection and Measurements
NIH —National Institutes of Health
NRC —Nuclear Regulatory Commission
NRC —National Research Council (National Academy of Sciences)
NTP —National Toxicology Program (DHHS)
OSHA —Occupational Safety and Health Administration (Department of Labor)
OTA —Office of Technology Assessment (U.S. Congress)
PBL —Peripheral Blood Lymphocyte
PFGE —Pulsed Field Gel Electrophoresis
PKU —Phenylketonuria
R —Roentgen
RFLP —Restriction Fragment Length Polymorphism
RNA —Ribonucleic Acid
SCE —Sister-chromatid Exchange
SLT —Specific Locus Test
T —Thymine
TG^r —Thioguanine resistant
TSCA —Toxic Substances Control Act
UNSCEAR—United Nations Scientific Committee on the Effects of Atomic Radiation
2DDGGE —Two Dimensional Denaturing Gel Electrophoresis
2DPAGE —Two Dimensional Polyacrylamide Gel Electrophoresis
6TG —6Thioguanine
$6TG^r$ —6Thioguanine resistant

Glossary of Terms

Achondroplasia: A disease marked by a defect in the formation of cartilage at the ends of long bones (femur, humerus) that leads to a type of dwarfism. There are a number of hereditary forms, the most common of which is autosomal dominant.

Allele: An alternative form of a gene, or a group of functionally related genes, located at the corresponding site on a homologous chromosome. Each allele is inherited separately from each parent. Alleles can be dominant, recessive, or co-dominant for a particular trait.

Alpha thalassemia: A genetic defect caused by an alteration in a portion of the gene coding for the alpha globin molecule. The result is an insufficient number of alpha globin molecules and a deficiency of adult hemoglobin.

Amino acid: One of a group of 20 molecules that bind together in various sequences to form all protein molecules. The specific sequence of amino acids determines the structure and function of a protein.

Amniocentesis: A procedure that involves withdrawing a sample (usually 2 to 8 milliliters) of the amniotic fluid surrounding the fetus in utero. This fluid contains cells shed by the developing fetus. The cells can be grown in cell culture and analyzed either biochemically or cytogenetically to detect a variety of genetic abnormalities in the fetus, including genetic diseases such as Down syndrome and Tay Sachs disease.

Autoradiography: A technique for identifying radioactively-labeled molecules or fragments of molecules, useful for analyzing DNA for the presence of mutations.

Autosome: A chromosome not involved in sex-determination. In a complete set of human chromosomes, there are 44 autosomes (22 pairs of homologous chromosomes) and a pair of sex-determining chromosomes.

Base pair (of nucleic acids): A pair of hydrogen-bonded nitrogenous bases (one purine and one pyrimidine) that join the two strands of the DNA double helix. Adenine pairs with thymine, and cytosine pairs with guanine.

Beta globin: A constituent of hemoglobin, which is a molecule consisting of four globin subunits and a heme group (two alpha and two beta globins).

Carcinogen: A chemical or physical agent that causes cancer.

Cell: The smallest membrane-bound protoplasmic body, consisting of a nucleus and its surrounding cytoplasm, capable of independent reproduction.

Cell culture: Growth in the laboratory of cells isolated from multicellular organisms. Each culture is usually of one cell type (e.g., lymphocytes, fibroblasts, etc.).

Cell line: A sample of cells, having undergone the process of adaptation to artificial laboratory cultivation, that is now capable of sustaining continuous, long-term growth in culture.

Chromosome: A threadlike structure that carries genetic information arranged in a linear sequence. It consists of a complex of nucleic acids and proteins.

Chromosome abnormalities: A group of pathological conditions associated with abnormalities in the number or structure (e.g., insertions, deletions, rearrangements) of chromosomes.

Chromosome banding: The chemical process of staining chromosomes to identify each pair of homologous chromosomes. Staining by various techniques produces patterns of light and dark bands (visible under the microscope) that are characteristic of each chromosome pair. Chromosome banding is particularly useful in detecting structural chromosome abnormalities such as inversions and deletions.

Codon: A sequence of 3 adjacent nucleotides in mRNA that specifies 1 of the 20 amino acids or the initiation or termination of the peptide chain. The linear sequence of codons determines the order in which amino acids are added to a polypeptide chain during translation.

Complementary DNA (cDNA): Single-stranded DNA that is synthesized from a messenger RNA template. It is often used as a probe to help locate a specific gene in an organism.

Congenital: Refers to a condition that is present at birth.

Congenital abnormality: Any abnormality, genetic or nongenetic, that is present at birth.

Cytoplasm: The contents of a cell exclusive of the nucleus. It consists of an aqueous solution and the organelles suspended in it and is the site of most chemical activities of the cell, including protein synthesis.

Denaturation: The separation of double-stranded DNA into its complementary, separate strands by treatment with chemicals or heat. The narrow range of temperature or chemical concentration at which DNA denaturation occurs is characteristic of the nucleotide sequence of the particular molecule.

Diploid: The chromosome state in which each homologous chromosome is present in pairs. All human somatic cells are diploid (i.e., they have 46 chromosomes), whereas reproductive cells, with 23 chromosomes, are haploid.

DNA: Deoxyribonucleic acid. The nucleic acid in chromosomes that contains the genetic information. The molecule is double-stranded, with an external "backbone" formed by a chain of alternating phosphate and sugar (deoxyribose) units and an internal ladder-like structure formed by nucleotide base-pairs held together by hydrogen bonds. The nucleotide base pairs consist of the bases adenine (A), cytosine (C), guanine (G), and thymine (T), whose structures are such that A can hydrogen bond with T, and C with G. The sequence of each individual strand can be deduced by knowing that of its partner. This complementarity is the key to the information-transmitting capabilities of DNA. (See also *nucleotide* and *gene*.)

DNA probe: A segment of complementary DNA that is used to detect the presence of a particular nucleotide base sequence.

DNA sequence: The order of nucleotide bases in DNA.

Dominant: A term used to refer to a genetic trait that is expressed in an individual who is heterozygous for a particular gene. Compare *recessive*.

Down syndrome: A genetic abnormality characterized by mental retardation, congenital heart defects, immune system abnormalities, various morphological abnormalities and a reduced life expectancy. Down syndrome is caused by either an extra copy of chromosome 21 (called trisomy 21) or by two copies of chromosome 21 and another chromosome 21 translocated to a different chromosome (usually to chromosome 14). The latter results in "translocation Down syndrome." Trisomy 21 has been shown to increase in frequency with advanced maternal age.

Electrophoresis: A technique used to separate molecules (such as DNA fragments or proteins) from a mixture of similar molecules. By passing an electric current through a medium containing the mixture, each type of molecule travels through the medium at a rate that corresponds to its electric charge and size. Separation is based on differences in net electrical charge and in size or arrangement of the molecule.

Enzyme: A protein that catalyzes a chemical reaction without being permanently altered or consumed by the reaction, so that it can be used repeatedly.

Enzyme deficiency variants: Abnormal proteins characterized by reduction or elimination of enzymatic activity. These changes may be caused by the absence of a gene product, or the presence of proteins that are non-functional or abnormally unstable.

Epidemiologic studies: Studies concerned with the relationships of various factors determining the frequency and distribution of diseases in a human population.

Escherichia coli (E. coli): A species of rod-shaped, gram negative bacteria that inhabit the normal intestinal tract of vertebrates. Many nonpathogenic strains of E. coli are hosts in recombinant DNA technologies.

Eukaryote: Cells or organisms with membrane-bound, structurally discrete nuclei and well-developed cell organelles. Eukaryotes include all organisms except viruses, bacteria, and blue-green algae. Compare *prokaryote*.

Exons: DNA sequences that determine the sequence of amino acids in proteins. Exons are separated on the DNA by introns, or intervening sequences, that are transcribed and later removed, or spliced out, during the production of mature messenger RNA.

Gametes: Mature male or female reproductive cells (germ cells: sperm or ova) with a haploid chromosome content (23 chromosomes in humans).

Gene: A linear sequence of nucleotides in DNA that is needed to synthesize a protein and/or regulate cell functions. A mutation in one or more of the nucleotides in a gene may lead to abnormalities in the structure of the gene product or in the amount of gene product synthesized.

Genetic code: The sequence of nucleotide triplets along the DNA that determines the sequence of amino acids in protein synthesis. This code is common to nearly all living organisms.

Genome: A term used to refer to all the genetic material carried by a single gamete.

Genotype: The genetic constitution of an organism, as distinguished from its physical appearance (its phenotype). For example, an individual may have a heterozygous genotype for eye color consisting of an allele for brown eyes (which is dominant) and an allele for blue eyes (which is recessive) or a homozygous genotype, with two alleles (both dominant) for brown eyes. In either case, the phenotype is the same: brown eyes.

Germinal mutations: Mutations in the DNA of reproductive cells—egg or sperm. Germinal mutations can be transmitted to the offspring only if one of those particular germ cells is involved in fertilization.

Germ line: The tissue or cell line that produces gametes and is used for reproductive purposes, as opposed to tissue or cell line from somatic cells. Also known as "germinal tissue."

Haploid: Half of the full set of genetic material, or one chromosome of each homologous pair. Gametes have a haploid set of DNA. Fertilization of an ovum by a sperm produces a diploid number of chromosomes in the zygote.

Hemoglobin: The oxygen-carrying molecule found in red blood cells. Adult hemoglobin is a protein composed of a single heme group bound to two alpha globin chains and two beta globin chains.

Hemoglobinopathies: A collection of hereditary disorders of hemoglobin structure and/or function. Examples are thalassemia and sickle cell anemia.

Hemophilia: A genetic disorder distinguished by a deficiency of one or more coagulation factors—e.g., Factor VIII (hemophilia A) or Factor IX (hemophilia B). The underlying mutations are located on the X chromosome. Hemophilia occurs most often in males who have only one X chromosome, and is transmitted to offspring by asymptomatic females, who have two X chromosomes, one of which carries the hemophilia mutation.

Heritable mutation: A mutation that is passed from parent to offspring and was present in a germ cell of one of the parents.

Heteroduplex: A double-stranded nucleic acid molecule composed of individual strands of different origin, e.g., a parent's DNA strand hybridized to a child's DNA strand, or RNA hybridized to DNA.

Heterozygote: An individual who has two different alleles of any one particular gene, e.g., an individual who has one copy of the gene for thalassemia at the locus for beta globin is heterozygous for this trait.

Homoduplex: A double-stranded nucleic acid molecule composed of two strands of the same origin, e.g., genomic DNA isolated from human cells (DNA hybridized with DNA from the same individual).

Homozygote: An individual with the same allele at both genes responsible for a particular trait.

Homozygous: Having identical alleles at a given locus. Compare *heterozygous*.

Hybridization: The process of joining two single-strands of RNA or DNA together so that they become a double-stranded molecule. For hybridization to occur, the two strands must be nearly or perfectly complementary in the sequence of the nucleotide base pairs.

Induced mutation: A change in the structure of DNA or the number of chromosomes, caused by exposure of the DNA to a mutagenic agent.

In vitro: Literally, "in glass," pertaining to a biological process or reaction taking place in an artificial environment, usually a laboratory. Sometimes used to include the growth of cells from multicellular organisms under cell culture conditions.

In vivo: Literally, "in the living," pertaining to a biological process or reaction taking place in a living cell or organism.

Karyotype: A photomicrograph of an individual's chromosomes arranged in a standard format, showing the number, size, and shape of each chromosome.

Kilobase: A unit of measurement for the length of nucleic acids along a chromosome. One kilobase (abbreviated kb) consists of 1,000 nucleotide bases. Genes can be several kilobases in length.

Lambda: A bacterial virus that infects E. coli, producing many copies of itself within the bacterium.

Lesch-Nyhan syndrome: An X-linked recessive disorder characterized by compulsive self-mutilation and other mental and behavioral abnormalities. It is caused by a defect in the gene that produces a particular enzyme (hypoxanthine-guanine phosphoribosyl transferase) important in metabolism. The causal relationship to the behavioral aspects of the disorder is not yet understood.

Locus: The position of a gene or of a group of functionally related genes on a chromosome.

Locus Test: A measurement of genes examined in a study of mutant proteins, based on the number of proteins examined, on the number of gene loci represented by each such protein, and on the number of individual samples obtained.

Lymphocytes: Specialized white blood cells involved in the body's immune response. B-lymphocytes originate in the bone marrow and when stimulated by an antigen produce circulating antibodies (humoral immunity). T-lymphocytes are produced in the bone marrow and mature in the thymus gland and engage in a type of defense that does not depend directly on antibody attack (cell-mediated immunity).

Marfan's syndrome: A genetic disorder characterized by abnormally long fingers and toes and by abnormalities of the eye lenses and heart. (Abraham Lincoln is thought by some to have suffered from this disease). Also called arachnodactyly ("spider fingeredness").

Meiosis: The process of cell division in reproductive cells that reduces the number of chromosomes to half of the full set so that each parent contributes half of the genetic material to each offspring. When the egg and sperm fuse at conception, the full set of chromosomes (a total of 46 chromosomes in humans) is restored in the offspring.

Mendelian: A term used to refer to a trait that follows Mendel's laws of inheritance and is controlled by a single gene, and which therefore shows a simple pattern of inheritance (dominant or recessive). So named because traits of this sort were first recognized by Gregor Mendel, the Austrian monk whose early research laid the basis for modern genetics. Mendel's laws include the Law of Segregation, which describes how each pair of alleles separates into different gametes and the Law of Independent Assortment, which describes how different alleles are assorted independently of the other alleles in gametes and how the subsequent pairing of male and female gametes occurs randomly.

Messenger RNA (mRNA): Ribonucleic acid that serves as the template for protein synthesis. It is produced by transcribing a DNA sequence into a complementary RNA sequence. Messenger RNA molecules carry the instructions for assembling proteins on the ribosomes.

Mitosis: The process of division involving DNA replication, and resulting in two daughter cells with the same number of chromosomes and cytoplasmic material as the parent cell.

Multifactorial traits: Traits that are not determined solely by a single gene. Multifactorial, or polygenic, traits (e.g., height) have variable phenotypic effects that depend for their expression on the interaction of many genes and environmental influences.

Mutagen: A chemical or physical agent that causes changes in the structure of DNA.

Mutation rate: The number of mutations per unit of DNA (e.g., per gene, per nucleotide, per genome, etc.) occurring per unit of time (usually per generation).

Mutations: Changes in the composition of DNA, generally divided according to size into "gene mutations" (changes within a single gene, such as nucleotide substitutions) and "chromosome mutations" (affecting larger portions of the chromosome, or the loss or addition of an entire chromosome). A "heritable mutation" is a mutation that is passed from parent to offspring and therefore was present in the germ cell of one of the parents. See also *induced mutation* and *spontaneous mutation*.

Nucleic acids: Macromolecules composed of sequences of nucleotides that carry genetic information. There are two kinds of nucleic acids: DNA, containing the sugar deoxyribose and RNA, containing the sugar ribose.

Nucleotide: A subunit of DNA or RNA, consisting of a nitrogenous base (adenine, guanine, thymine, cytosine, or uridine), a phosphate molecule, and a sugar molecule (deoxyribose in DNA or ribose in RNA). The linkage of thousands of these subunits forms the DNA or RNA molecule.

Nucleus: The intracellular structure of eukaryotes that contains DNA.

Oligonucleotide: A polymer made up of a few (generally fewer than 10 or 20) nucleotides; a short sequence of DNA or RNA.

Peptide: A compound consisting of two or more amino acids linked together, formed through a chemical process that produces one molecule of water for each joining of one amino acid to another. Peptides are the building blocks of proteins.

Phenotype: The appearance of an individual or the observable properties of an organism that result from the interaction of genes and the environment.

Phenylketonuria (PKU): An autosomal recessive genetic disorder of amino acid metabolism, caused by the inability to metabolize phenylalanine to tyrosine. The resulting accumulation of phenylalanine and derived products causes mental retardation, which can be avoided by dietary restriction of phenylalanine beginning soon after birth.

Polymorphism: A single gene trait (e.g., a red blood cell surface antigen) that exists in two or more alternative forms (such as types A, B, AB, and O blood). Genetic variants are considered polymorphisms if their frequency exceeds 1 percent each, but are considered "rare mutations" if they are found in less than 1 percent of the population.

Polypeptide: A sequence of amino acids joined in a chain.

Probe: A molecule that has been tagged or labeled in some way with a tracer substance (a radioactive isotope or specific dye-absorbing compound) that is used to locate or identify a specific gene or gene product. For example, a radioactive mRNA probe for a DNA gene, or a monoclonal antibody probe for a specific protein. See also *DNA probes*.

Prokaryote: A cell or organism lacking membrane-bound, structurally discrete nuclei. Compare *eukaryote*.

Protein: A molecule composed of hundreds of linked amino acids in a specific sequence, which is, in turn, determined by the sequence of nucleotides in DNA. Proteins are required for the structure, function, and regulation of cells, tissues, and organs in the body.

Recessive: A genetic trait that is manifested phenotypically only when both alleles for the trait are present at a locus. E.g., sickle cell anemia is manifest only when both copies of the gene for beta globin contain the $beta^s$ mutation, whereas when individuals have only one copy of the $beta^s$ gene, they do not develop the disease, but have sickle-cell trait, a generally benign condition. X-linked traits act as if they were recessive in females and dominant in males. Compare *dominant*.

Recombinant DNA (rDNA) technology: Techniques involving the incorporation of DNA fragments, generated with the use of restriction enzymes, into a suitable host organism's DNA (a vector). The host is then grown in culture to produce clones with multiple copies of the incorporated DNA fragment. The clones containing this particular DNA fragment can then be selected and harvested.

Replication: The synthesis of new DNA from existing DNA.

Restriction enzyme: An enzyme that has the ability to recognize a specific nucleotide sequence in a nucleic acid (ranging from 4 to 12 base pairs in length) and cut or cleave the nucleic acid at that sequence. They are termed "restriction" enzymes because, occurring naturally in bacteria, they recognize foreign nucleic acid (e.g., the DNA of a bacterial virus as it begins to infect and destroy its host) and destroy it, thus restricting the ability of the virus to prey upon certain potential host strains. Over 400 different restriction enzymes are known, recognizing a great variety of different nucleotide base sequences. Use of restriction enzymes has made possible the cutting and splicing together of nucleic acid within and between different organisms and species.

Restriction Fragment Length Polymorphism (RFLP): Fragments of DNA that may vary in length from individual to individual, depending on the presence of polymorphisms or rare mutations.

Ribosome: A cellular organelle that is the site of protein synthesis, the process of reading the instructions in mRNA and using it as the guide to constructing the specified protein.

RNA (ribonucleic acid): The nucleic acid found mainly in the nucleus and ribosomes and is involved in the control of cellular activities. There are several classes of RNA that serve different purposes, including messenger RNA (mRNA), transfer RNA (tRNA), and ribosomal RNA (rRNA). The functions of various kinds of RNA include "translating" selected gene sequences coded in the cell's DNA, and transferring that information outside of the nucleus to structures that then synthesize the proteins indicated by the RNA vehicle.

RNaseA: An enzyme that cleaves RNA where certain sequences of nucleotide bases occur. RNaseA has been used to cut fragments of RNA/DNA heteroduplexes where particular mismatches in the heteroduplexes occur.

Sentinel phenotypes: A group of autosomal dominant or X-linked conditions that can occur sporadically. They are thought to result from new germinal mutations in the parents' reproductive cells.

Sex chromosomes: The X and the Y chromosomes, 2 of the 46 chromosomes in human cells, that determine the sex of the individual. Females have two X chromosomes, while males have one X and one Y chromosome.

Sickle-cell disease: A genetic disorder of hemoglobin function caused by the presence of an abnormal beta globin chain. Patients with sickle-cell disease have red blood cells that tend to deform into a sickle-like shape. The specific defect is caused by an abnormal gene, resulting in the replacement of the usual amino acid, glutamic acid, with valine, in the sixth amino acid position in the beta-hemoglobin molecule. This increases its propensity to crystallize, thus rupturing the red blood cell and causing the cells to lodge in small blood vessels. Also called "sickle-cell anemia." See *sickle-cell trait*.

Sickle-cell trait: The generally benign condition shown

by individuals carrying the variant beta[s] gene as well as the normal beta[a] gene. Such individuals are heterozygous for the sickle-cell gene, and are healthy (i.e., are usually asymptomatic), but two heterozygous parents have a 25 percent risk with each pregnancy of having a child with *homozygous* sickle-cell disease.

Single gene disorder: A genetic disease caused by a single gene and showing a simple pattern of inheritance (e.g., dominant or recessive, autosomal or X-linked). Also called "Mendelian disorder."

Somatic: A term used to refer to body tissues, as opposed to reproductive (germinal) tissues.

Somatic cell: Any cell in the body except reproductive cells or their precursors.

Spontaneous mutation: In the absence of any known causative agent, a change in the structure of DNA or in the number of chromosomes. Also called a "background" mutation. Also see *mutations*.

Stem cells: Undifferentiated cells in the bone marrow that have the ability to replicate and to differentiate into blood cells.

Tay-Sachs disease: An autosomal recessive genetic disease resulting in developmental retardation, paralysis, dementia and blindness, usually fatal in early childhood. The defective gene codes for hexosaminidase A, an enzyme that is involved in certain chemical pathways in the brain. Symptoms are caused by an accumulation of cerebral gangliosides, fatty acid and sugar molecules found in the brain and nervous tissue. The gene is found in highest frequency among Jews of Eastern European origin.

Teratogen: A physical or chemical agent (e.g., thalidomide, radiation, alcohol, etc.) that can cause congenital abnormalities as a result of exposure in utero.

Tetramer: A complex molecule consisting of four major portions joined together (e.g., hemoglobin, in which two alpha chains and two beta chains are joined to a central heme group.)

Thalassemia: A group of autosomal recessive genetic disorders characterized by abnormalities in synthesis of the globin polypeptides of hemoglobin. The two most common forms are alpha-thalassemia and beta-thalassemia, disorders of the alpha- and beta-globin polypeptides, respectively, which cause imbalances in the production of the globins and lead to an overall deficiency of adult hemoglobin. The thalassemias are most common in people of Mediterranean, Middle Eastern, and Asian descent.

T-lymphocytes: See *lymphocytes*.

Transcription: The synthesis of messenger RNA (mRNA) on a DNA template. The resulting RNA sequence is complementary to the DNA sequence.

Transfer RNA (tRNA): Specialized RNA molecules that function to bring specific amino acids to ribosomes that translate messenger RNA (mRNA) into proteins.

Translation: The process in which the genetic code contained in the nucleotide base sequence of messenger RNA directs the synthesis of a specific order of amino acids to produce a protein.

Trisomy: The presence of an extra chromosome, resulting in three homologous chromosomes instead of two, e.g., Down syndrome can result from Trisomy 21, or the presence of an extra chromosome number 21 in each body cell.

tRNA: See *transfer RNA*.

X-linked mutation: A mutation that occurs in a region of the X-chromosome.

Zygote: A fertilized egg that results from the fusion of sperm and egg.

Chapter 1

Summary and Options

Contents

FIGURE

Chapter 1

Summary and Options

INTRODUCTION

Mutations, lasting changes in the genetic information carried in the deoxyribonucleic acid (DNA) of cells, can cause severe diseases and disabilities, none of which is curable and relatively few of which can be treated effectively. Such genetic diseases represent a significant fraction of chronic disease and mortality in infancy and childhood; they generally impose heavy burdens expressed in premature mortality, morbidity, infertility, and physical and mental handicap. Some of the most common of the 3,000 or more different disorders known to result from mutations include Down syndrome, Duchenne muscular dystrophy, and hemophilia. In addition, mutations have been associated with increased susceptibilities to certain chronic diseases, including some forms of diabetes, heart disease, and cancer. Most mutations that are expressed as genetic disease already exist in the population and are carried from generation to generation. A smaller proportion of mutations arises anew, "sporadically," in each generation, and the specific causes of these mutations are unknown.

The public and the government have expressed concern about the possibility that environmental exposures are contributing to or increasing the frequency of mutations. Mutations are among the chronic health effects singled out in the Toxic Substances Control Act of 1976 (TSCA) and in the Comprehensive Environmental Response, Compensation, and Liability Act of 1980 (CERCLA or "Superfund"). In those laws, Congress specified public protection from exposures that can cause mutations. The other major environmental statutes contain language broad enough to include protection from mutagens.

Unfortunately, little is currently known about the kinds and rates of mutations that occur in human beings. Available methods to study such mutations are inadequate to provide sufficient information for evaluating mutagenic risks.

Much of our knowledge of genetic risks to human health from exposures to environmental agents has been derived from the study of the effects of mutagens on experimental animals. These experiments are useful in manipulating various aspects of the mutagenic process, for example, to examine how mutagens act on DNA and to study effects of varying doses and rates of exposure to mutagenic agents administered either singly or in combination. Experimentation with animals is essential for assessing potential hazards of new chemical and physical agents before human populations have been exposed to them. However, at present, technical problems in detecting and measuring mutations limit animal experiments as they limit human studies, so the results from animal experiments, using current methods, detect only a small proportion of the kinds and numbers of mutations that can occur.

Recent advances in molecular biology have led to the development of new technologies for examining DNA that may provide insight into the kinds and rates of mutations that occur in human beings. This report assesses these developments and discusses their potential for predicting risks of mutation from particular exposures.

At present, these new technologies propose reasonable and verifiable ways of detecting heritable mutations in human DNA and proteins, but none is efficient enough to be used on a large scale. However, there is considerable optimism in the scientific community that these new technologies can provide, for the first time, the means to obtain basic knowledge about the primary causes of mutation and the means to assess the kinds and rates of mutations that occur in human beings.

Data derived from studies in human beings, along with verifiable methods to extrapolate from corresponding animal data, will permit a more informed assessment of the medical and biological consequences of mutagenic exposures. At present, without such comparative data, it is difficult to know whether general extrapolations from animal data would lead to underestimates or overestimates of the genetic risks for humans. Continuing to rely on inadequate data may incur both human and financial costs, since conclusions drawn from this information contribute to decisions about acceptable levels of exposure and the level of society's resources that are devoted to providing protection from such exposures.

A combination of factors—concern that environmental exposures may be contributing to human mutations, questions about the fundamental nature of mutations, and increasing knowledge of the structure and function of DNA—increase the likelihood that new technologies will be developed and field tested. However, studies using these technologies may be expensive and will probably require the collaboration of a large number of scientists; their continued development, pilot testing, and large-scale application may require sufficient interest and financial support outside the scientific community. With such support, and with continued development of the techniques, some of these techniques could be ready for large-scale use in the next 5 to 10 years.

Congressional interest in supporting basic research on human mutations and in the continued development of these technologies is necessary if the regulatory agencies are eventually to have the tools to evaluate risks from most occupational or environmental exposures. The current lack of information on kinds and rates of human mutations is largely due to the inadequacy of present methods to study heritable mutations. Efforts to comply with the agencies' mandates to protect people from mutagens may be impeded unless basic knowledge of causes, kinds, and rates of human mutations is obtained.

Request for the Assessment

This assessment was requested by the Senate Committee on Veterans' Affairs, the House Committee on Science and Technology, and the House Committee on Energy and Commerce. Interest in the assessment was also expressed by the Senate Committee on Public Works and the Environment, the Senate Committee on Labor and Human Resources, and the House Committee on Veterans' Affairs.

These committees have wrestled with problems of determining whether past exposures to potential mutagens have affected the health of veterans and civilians and of framing reasonable public health laws that can be implemented, given current knowledge and technologies. OTA was asked to assess the available information about current means for detecting mutations as they relate to these issues and on the likelihood and potential impact of technological developments.

Scope of the Report

This chapter summarizes current knowledge about the kinds and rates of human mutations and the methods that have been used to detect heritable mutations in human beings and in experimental animals. New technologies assessed in this report for detecting and measuring human heritable mutations are briefly described. Methods for measuring human somatic mutations are discussed as tools for evaluating the risks of heritable mutations. The final section of this chapter presents options for congressional action.

Chapter 2 provides background information about human genetics and DNA, and discusses the types of mutations that can occur and their potential health effects. Chapter 3 reviews the literature on current methods for studying mutations and summarizes current knowledge about the frequency of heritable mutations in human populations.

The new technologies for examining human DNA for heritable mutations are described in chapter 4, followed by descriptions of new somatic mutation tests in chapter 5.

Chapter 6 summarizes data from experimental animals on spontaneous and induced mutations, and discusses the possible use of such data for identifying human mutagens and determining their potency. Chapter 7 focuses on the problems of extrapolating from the results of animal experiments to human risks.

Chapter 8 discusses epidemiologic considerations in the application of the new technologies, such as validation of the new methods and selection of at-risk populations to study. Chapter 9 discusses Federal involvement in protecting against genetic risks and the regulatory mechanisms available to control exposures to mutagenic agents.

A summary of current Federal expenditures in mutation research and potential costs of studies to detect mutations using the new technologies is presented in appendix A.

BACKGROUND

Kinds and Effects of Mutations in Human Beings

Mutations can occur "spontaneously," that is, in the apparent absence of any unusual stimuli, or they can be "induced" by particular agents. It is likely that many or most "spontaneous" mutations are caused by external forces, possibly including ionizing radiation, ultraviolet radiation, viruses, and certain chemicals, but the appropriate links have not been made. Some mutagens present around us may also be necessary for sustaining life, for example, oxygen, components of our food, and some of the body's own metabolites. Experiments in animals have shown that many substances present in agricultural, industrial, and pharmaceutical chemicals in use today are mutagenic in some test systems. Which of these cause mutations in human beings is still a matter of speculation. Precise causes for essentially all mutations that have been identified in human beings are unknown.

At present, more than 3,000 different genetic diseases have been identified, including disorders resulting from mutations in DNA, and disorders resulting from the interaction of genetic and environmental components. Approximately 10 in 1,000 liveborn infants are born with a single gene disorder and an additional 6 in 1,000 liveborn infants are born with a major chromosome abnormality. It is estimated that approximately 80 percent of the single gene disorders are the direct result of mutations that occurred in germ cells of distant ancestors and were passed along to succeeding generations. The remaining 20 percent of these cases (0.2 percent of all livebirths) and the majority of chromosome abnormalities are believed to be due to sporadic mutations in the reproductive cells of one of the parents of the infant. An additional 10 in 1,000 liveborn infants manifest a serious genetic disease sometime after birth, and a far higher proportion of newborns will show indirect effects of one or several parental or ancestral mutations in later life as, for example, in increased susceptibilities to some forms of heart disease, diabetes, or cancer.

Mutations are changes in the composition of the genetic material, DNA (see fig. 1), and are generally divided according to size into gene mutations and chromosome mutations. Gene mutations refer to changes within a single gene, for example, substitutions of single component nucleotides, or small losses or additions of genetic material in expressed or nonexpressed regions of the gene. Chromosome mutations affect larger portions of the chromosome (e.g., structural rearrangements of genetic material in the chromosomes) or result in the loss or addition of an entire chromosome. Since DNA directs the synthesis and regulation of molecules in the body, either group of mutations can influence a wide range of biological and physiological functions, including reproduction, longevity, intelligence, and physical development. Individual differences in susceptibility to disease may result from the effects of one gene, several genes, or combinations of genes and environmental factors.

Depending on the nature and location of the mutations and on the function of the genes in which they occur, mutations may, in theory, be beneficial, neutral, or harmful to the individual.[1] The kinds and effects of known mutational events range from single nucleotide substitutions (the smallest unit of change in DNA) with no clinically observable effects, to single nucleotide sub-

[1]Current theory maintains that *most* newly arising mutations in regions of DNA that directly determine the structure and regulation of proteins are more likely to be detrimental than beneficial. Little is known of effects of mutations in other regions of DNA.

Figure 1.—Organizational Hierarchy of DNA, the Carrier of Genetic Information in Human Cells

A G C T

Nucleotides

Adenine, guanine, cytosine, and thymine, the basic building blocks of DNA.

AAA CGC GAC CGA

Codons

Nucleotides arranged in a triplet code, each corresponding to an amino acid (components of proteins) or to a regulatory signal.

-ACGAAAATCCGCGCTTCAGATACCTTA-

Genes

Functional units of DNA needed to synthesize proteins or regulate cell function.

Chromosome

Thousands of genes arranged in a linear sequence, consisting of a complex of DNA and proteins.

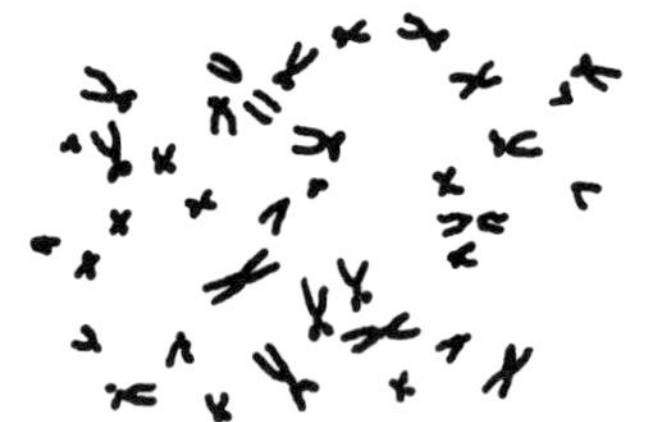

Genome

The complete set of genetic information; each human reproductive cell contains 23 chromosomes, and all other cells in the body contain a full set of 46 chromosomes.

SOURCE: Office of Technology Assessment.

stitutions resulting in severe diseases; from major structural and numerical chromosome abnormalities (the largest observed unit of change in DNA) leading to various abnormalities and impairments, to those resulting in embryonic, fetal, or neonatal death.

A child's entire genetic endowment comes from the DNA of two single reproductive cells (or gametes), one egg and one sperm, from his parents.[2] A mutation occurring in the DNA of either of these germ cells, a *germinal mutation*, is passed on to the child, who is born with a "new" *heritable mutation*. Mutations in germ cells that are not involved in fertilization are not passed on to the offspring. If a mutation arises in the DNA of the parents' nonreproductive cells (collectively termed somatic cells), such *somatic mutations* are not transferred to the reproductive cells. Somatic mutations may, however, affect the parents' health, and indirectly, their ability to bear a healthy child.

[2]Each of the gametes (the male spermatozoan and the female ovum) is the product of a series of developmental stages of reproductive (or germ) cells.

CURRENT METHODS FOR STUDYING MUTATIONS

Current empirical methods to study mutations in human beings focus on physiological and biochemical effects of mutations because, until recently, it was not possible to examine changes in DNA directly. Each of the current methods detects only a limited portion of the spectrum of mutational changes. These methods have been used to derive estimates of baseline frequencies of some kinds of human mutations.

Animal Studies

Much of our knowledge about how substances interact with DNA and how they may cause mutations is derived from studies with experimental animals. In addition, some estimates of human mutation rates have been derived, by extrapolation, from animal studies. Beginning soon after World War II and still continuing, spontaneous and induced gene mutation rates have been studied in laboratory mice. The rate of spontaneous heritable gene mutations, as detected in these experiments, is roughly two to eight mutations per 1 million genes per generation of mice. Radiation and approximately two-thirds of about 20 chemicals[3] tested so far increase the frequency of de-

[3]On the basis of animal experiments using the specific locus test and the heritable translocation test, in which they were found to be mutagenic in all germ cell stages, these chemicals are strongly suspected to be mutagenic in humans (154). They included both common environmental agents and chemicals available in the laboratory that are not normally found in the environment.

tectable heritable gene mutations in these experimental mice. Generally, the chemical substances that induce heritable mutations in mature and maturing germ cells also induce somatic mutations, and vice versa. However, substances that induce somatic mutations do not necessarily induce heritable mutations in immature germ cells (stem cells). Unlike mature germ cells or somatic cells, these germinal stem cells may have efficient systems for repair of mutational or premutational damage. There is, in fact, indirect evidence from dose-response and dose-rate experiments that germinal stem cells have good repair systems.

It may be useful to quantify relationships between somatic and germinal cells with regard to mutagenic potency of different chemicals, and to study mutations in equivalent sets of genes in both types of cells. Animal experimentation is useful for determining the feasibility of using human somatic mutation rates for predicting the risk of human heritable mutations. This work in animals may demonstrate whether it is possible to extrapolate from the occurrence of somatic to that of heritable mutations at all, and may help determine whether it is possible to generalize from animal data to human beings. In addition, animal studies on heritable mutations are useful not only for determining whether a given agent is mutagenic but also for more general explorations of the factors that may influence the occurrence of mutations.

Studies in Human Beings

Spontaneous heritable mutations in human beings have been studied by examining: 1) the incidence of certain genetic diseases ("sentinel phenotypes"), 2) gross changes in chromosome structure or number, and 3) changes in the structure or function of blood proteins. Epidemiologic studies of specific populations, in particular, the survivors of atomic bombs in Japan, provide some information about induced heritable mutations in human beings.

Sentinel Phenotypes

The classic method for identifying human heritable mutations is the empirical observation of infants and children for the presence of certain rare genetic diseases known as sentinel phenotypes. Examples include achondroplasia (dwarfism), aniridia, and some childhood cancers, such as retinoblastoma and Wilms' tumor. By recording the occurrence of sentinel phenotypes as a proportion of the total number of livebirths in a defined population over time, the frequency of each disorder (and of its corresponding mutation) can be estimated (165).

The characteristic of sentinel phenotypes that is most useful for mutation studies is that these conditions are "sporadic" in most or all cases; they almost always result from a new germinal mutation in one of the parents of the afflicted individual.[4] Each different sentinel phenotype is thought to result from a different, single, mutant gene, although precise genetic information to confirm the single gene hypothesis is lacking in most cases.

Despite the distinctive characteristics of sentinel phenotypes, the relevance of existing data on the frequency of the various sentinel phenotypes to the study of kinds and rates of human mutations is limited by a lack of knowledge of the genetic bases of the phenotypes, and by the small fraction of DNA that accounts for these phenotypes. Of the several thousand known genetic diseases, only 40 are thought to satisfy the criteria for inclusion as "sentinel phenotypes." Roughly 40 genes are involved in the 40 sentinel phenotypes, among a total of an estimated 50,000 to 100,000 expressed genes in an individual's DNA.

Sentinel phenotypes are severely debilitating conditions that require accurate diagnosis and long-term medical care. However, practical difficulties arise in gathering and maintaining data on the incidence of sentinel phenotypes for the purpose of tracking mutation rates. Infants with sentinel phenotypes are rare, numbering approximately 1 in 10,000 to 1 in 10 million liveborn infants, depending on the particular disease. Consequently, a huge number of infants must be observed in order to find even a few infants with sentinel phenotypes. Diagnosis of individual phenotypes is complicated by the genetic heterogeneity of these disorders, so that highly trained specialists in various pediatric subdisciplines, which are

[4]Various tests are done to exclude nonsporadic cases, which could result from X-linked recessive inheritance, mistaken parentage, and the occurrence of other genetic or nongenetic conditions that mimic the appearance of sentinel phenotypes.

few in number, would be needed to make these observations. Millions of consecutive newborn infants would have to be monitored thoroughly and accurately for many years, and registries, much larger than those currently in use, would be needed to collect and maintain the necessary data.

Chromosome Abnormalities

Another method for identifying a certain class of human heritable mutations is the examination of chromosomes under a light microscope for the presence of chromosome abnormalities ("cytogenetic analysis"). Normal human DNA is organized into 46 chromosomes (44 autosomes and 2 sex chromosomes), distinguishable by size, proportional shape, and staining pattern, or "banding." Chromosome abnormalities are defined as either *numerical* (extra or missing whole chromosome[s]) or *structural* (deletions, insertions, translocations, inversions, etc., of sections of chromosomes). In total, chromosome abnormalities are estimated to occur in at least 5 percent of all human conceptions. The majority of such conceptuses are spontaneously aborted, but the few that survive comprise approximately 0.6 percent of all liveborn infants. The incidence of Down syndrome is 1 in 650 at birth, making it the most common chromosomal disorder in newborns.

With the most advanced chromosome staining methods currently available, approximately 1,000 bands are distinguishable in one set of human chromosomes, although far fewer bands are produced with routine methods. Mutations in DNA sometimes cause a change in the banding pattern, particularly if such mutations involve large sections of a chromosome. With routine banding methods, there may be several hundred genes present in each visible band, and with higher resolution banding methods, a single band may contain about 100 genes. However, smaller mutations, from single nucleotide changes within genes on up to some deletions and insertions of entire genes, generally are not visible by any banding method.

Measurement of Mutant Proteins

Most of the available information on rates of spontaneous human mutations has been derived from studies of sentinel phenotypes and chromosome abnormalities, but data on different kinds of mutations in humans is now emerging from studies of mutant proteins. In general, mutant proteins are more precise indicators of genetic damage than are clinical and cytogenetic observations. Certain mutations in genes that determine the structure of proteins alter the chemical characteristics of the proteins, causing them to behave differently in separation and purification procedures. These differences suggest that a mutation has occurred because proteins are constructed according to blueprints in DNA, and changes in DNA can lead to the production of altered proteins. If the protein under study is an enzyme, a mutation within the gene that codes for it can alter, diminish, or eliminate the enzyme's biochemical activity.

Operationally, mutant proteins are identified by taking samples of blood from each member of a "triad," including both parents and the child. Proteins are extracted from blood components, and the proteins are separated by electrophoresis. Putative mutations are identified when a protein from a child behaves differently from the corresponding protein from both parents.

One-Dimensional Separation of Proteins.—The technique most commonly used to study mutant proteins is electrophoresis, a method of separating proteins on the basis of their electrical charges. The term "electrophoretic variant," or "electromorph," is used to describe a protein that behaves differently in electrophoresis from the corresponding protein found in the parents.

Although one-dimensional electrophoretic analysis of proteins is well established and can be improved by including functional assays for additional enzymes, it is limited to detecting: 1) mutations that do not eliminate the functional ability of the proteins, 2) nucleotide substitutions only in coding regions of genes for the proteins examined, and 3) only those nucleotide substitutions that alter the electrical charge on these proteins; such substitutions are thought to account for about one-third of all nucleotide substitutions in coding regions, which, in turn, account for a fraction of all the kinds of mutations that can occur. Electrophoresis does not detect many other types of mutations, including small duplications, rearrangements, or mutations that result in the absence of gene products; these mutations are thought to constitute the majority of the mutations induced by certain mutagens, including radiation. Moreover, electrophoresis does not detect mutations that occur anywhere outside the coding regions of a certain set of genes, including mutations in other coding genes and in non-expressed regions of the DNA.

Data from several studies using one-dimensional electrophoresis have been used to estimate the rate at which mutations produce electrophoretic variants, and from this estimate, to infer the total rate of amino acid substitutions in proteins, and the corresponding mutation rate per nucleotide in human DNA.

Two-Dimensional Separation of Proteins.—An extension of one-dimensional electrophoresis involves separation of proteins in a second dimension. With two-dimensional electrophoresis, about 300 proteins from each person can be separated and examined, compared with about 100 proteins per sample that can be separated in one-dimensional electrophoresis. Further improvements may be possible with the use of computer algorithms to assist in interpreting the complex two-dimensional gels. This technique is currently feasible, and although it detects the same types of mutations as one-dimensional electrophoresis, it can examine more proteins per sample.

Epidemiologic Studies

An extensive body of data from experimental animals demonstrates that exposure to radiation and to certain chemicals can induce mutations in mammalian germ cells. In humans, exposure to ionizing radiation is known to cause somatic mutations, and it is suspected to enhance the probability of heritable mutations. *To date, however, the available methods have provided no direct evidence for the induction (by chemicals or by radiation) of mutations in human germ cells.*

The single largest population studied for induced mutations is the group of survivors of the atomic bombs detonated in Hiroshima and Nagasaki in 1945. Many survivors of the bombs received doses of radiation that could have caused germinal mutations; in experiments with mutation induction by radiation in mammals, similar kinds and doses of radiation were sufficient to cause observable mutations in offspring. Therefore, it was assumed that germinal mutations could have been induced in people exposed to the radiation from the blasts.

Medical examinations of the survivors soon after the blasts revealed the immediate effects of whole body irradiation: loss of hair; reduction in bone marrow activity; and reduction in circulating white blood cells, associated with a reduction in the body's resistance to infection. Among those who recovered from the immediate effects of the radiation, there was a significant excess of cancer deaths later in life. Certain types of leukemia were the first cancers to appear in excess, but continuing followup has revealed later increases in other cancers, such as multiple myeloma and cancers of the breast, thyroid, colon, esophagus, stomach, lung, ovaries, and possibly of the spinal cord and nerves (24,55).

Exposure of pregnant women to radiation was found to be associated with an increased incidence in their liveborn infants of small head circumference, mental retardation, and an increased incidence of childhood cancers. The critical time for fetal brain damage from radiation exposure was identified as the period of 8 to 15 weeks of gestation (99).

Analysis of the chromosomes prepared from peripheral blood lymphocytes of survivors exposed to the radiation has indicated an excess of chromosome aberrations (7). Certain types of chromosome aberrations (mainly balanced structural rearrangements, such as reciprocal translocations and inversions) have been found to persist in circulating lymphocytes long after exposure to radiation, whereas other types of chromosome aberrations (e.g., unbalanced rearrangements) in lymphocytes declined in number soon after exposure. Overall, the frequency of chromosomally aberrant cells in the survivors' blood was found to be proportional to the estimated dose of radiation received at the time of the bombing. It has not yet known whether these somatic mutations are correlated with specific cancers or other diseases in the survivors.

Survivors' children who were conceived after the acute radiation exposure were examined to study mutagenic effects on the parents' reproductive cells. Using various methods available from 1945 to the present, survivors' offspring were studied for "untoward pregnancy outcomes,"[5] for certain chromosome abnormalities, or most recently, for abnormal blood proteins. The offspring of parents exposed to atomic radiation were compared with the offspring of parents who were beyond the zone of radiation (greater than 2,500 meters from the hypocenter at the time of the bombings). Observation and analysis of some 70,000 offspring has revealed no statistically significant excess in the incidence of stillbirths, congenital malformations, neonatal deaths, or chromosome abnormalities. These findings suggest that the frequency of radiation-induced germinal mutations that led to certain gross abnormalities in newborn infants was not high enough to be detectable in a population of that size and genetic heterogeneity. However, they do not rule out the possibility of other manifestations of genetic damage in these children, or of latent expressions of such damage, since the methods used to study this population could examine only a small subset of DNA and only a limited number of genetic endpoints.

Analysis of the children's blood proteins for electrophoretic variants was later done to detect recessive mutations, that is, mutations that are not expressed as disease (unless present in both copies of a particular gene). This analysis, begun in 1976, found few mutations in either the exposed or control groups, making interpretation problematic. While the results indicate no significant excess of mutant proteins in the children of exposed parents, they do not exclude the possibility that an excess exists undetected. Unfortunately, electrophoresis is inefficient at detecting deletions, one of the most likely types of radiation-induced mutations. Overall, these findings do not rule out the possibility of genetic damage to the offspring of survivors of the atomic bombs, but they put upper limits of the frequency of occurrence of certain types of mutations that the current methods are able to detect.

Taking these findings at face value, and cognizant of an enormous body of data on the genetic effects of radiation on experimental animals, the investigators suggest that the dose of radiation necessary to double the human mutation rate (the "doubling dose") was between 139 and 258 rem,[6] but they caution that there is a possibility of large error attached to that estimate since genes other than the ones sampled may demonstrate different sensitivity to radiation, and some types of mutations may be repaired more efficiently than others. Their estimate of the doubling dose, if correct, however, indicates that man could be considerably *less* sensitive to radiation than laboratory mice.

[5]These were defined as major congenital defects and/or stillbirths and/or death in the survivors' offspring during the first postnatal week. These abnormalities can be caused by exposure to radiation, as well as to other environmental agents, and by socioeconomic factors.

[6]A rem (Roentgen-Equivalent-Man) is a measure of absorbed radiation dose. For comparison, a chest X-ray exposes an individual to about 0.1 rem.

NEW TECHNOLOGIES FOR DETECTING HUMAN MUTATIONS

Recent advances in molecular biology have led to the development of techniques that allow direct examination of DNA for evidence of mutations. These methods can examine large regions of human DNA without requiring detailed knowledge of the genes contained in those regions and without preparing genetic probes for particular sequences. Unlike current methods—observing sentinel phenotypes, chromosome abnormalities, and electrophoretic variants—that are limited to detecting a small fraction of all kinds and numbers of mutations, the new techniques have the potential for detecting a wide, unselected spectrum of mutations across the DNA. These techniques are examples of the state-of-the-art in molecular genetics and they are now promising to provide the basis for better approaches to studying mutagenesis.

Applying recent developments in molecular biology to the problem of detecting sporadic mutations, the new techniques described in this report propose reasonable and verifiable ways of examining human DNA for alterations in sequence and structure. These developments include the ability to clone specific genes, to cut up DNA into predictable fragments, to hybridize complementary DNA strands, to detect less-than-perfect hybridizations due to single base pair changes, and to separate large fragments of human DNA. Some of these methods detect similar types of mutations and some complement each other by detecting different types. Some of these may merit further development and, eventually, pilot testing. Several new technologies, representing different approaches, are discussed below.

None of these new techniques has been applied to large human populations or to experimental animals and it is not known how well they will perform. At present, none of these techniques is approaching the efficiency needed for examining the kinds and rates of mutations in a population or for determining whether mutation rates are increasing. With technical improvements in efficiency, some of the techniques, or derivatives of them, could be available in the next 5 to 10 years for large-scale use. Since these technologies provide new information about DNA, the health implications of any newfound mutations may not be immediately known. Additional research and methods would be needed to examine biological and physiological implications of the identified mutations for the populations studied and for their descendants.

Detection and Measurement of Heritable Mutations

Restriction Fragment Length Polymorphisms

Restriction endonucleases are enzymes, isolated from bacteria, that can be used experimentally to cleave isolated DNA molecules into fragments at specific sites in the DNA sequence that they recognize. If any of these sites have been altered by mutation, the resulting pattern of fragment sizes would also be altered. Restriction enzymes can be used to detect mutations that either: 1) create a new restriction site, 2) eliminate an old one, or 3) change the distance between existing restriction sites. (These mutations may include single nucleotide substitutions, or multiple nucleotide deletions or insertions.)

To use this method to detect mutations, DNA would be treated with a set of restriction endonucleases and the resulting DNA fragments separated by electrophoresis and examined for differences between those present in the child's DNA and those in either parent's DNA. Restriction site analysis does not allow examination of every nucleotide. However, the use of a set of combined restriction enzymes increases the number of restriction sites identified, allowing a larger portion of the DNA to be examined, including both expressed and nonexpressed regions.

Genomic Sequencing

The most straightforward approach to looking for mutations is by determining the sequence of every nucleotide in a child's genome and then comparing this with the DNA sequences of the child's parents. To determine its sequence, human DNA is cut with restriction endonucleases into fragments, and then each fragment is analyzed for its sequence of nucleotides. Genomic sequencing would detect mutations regardless of where they occur—in regions that code for specific proteins and regulatory functions as well as in regions without known functions—and is therefore potentially very informative. While it is technically possible at present, sequencing is currently feasible only for very limited sections of DNA, such as the length of DNA comprising only a few genes. Because of current technical inefficiencies, it would be an enormous task, involving many laboratories, a large number of scientists, and at least several decades to sequence even one entire genome, the complete set of DNA in an individual's germ cell. At present, it is not feasible to use genomic sequencing to examine several peoples' genomes for mutations, although sequencing can be used in conjunction with other techniques to examine small sections of DNA.

One-Dimensional Denaturing Gradient Gel Electrophoresis

A modification of the standard electrophoretic gel procedure, denaturing gel electrophoresis allows DNA fragments to be separated not only on the basis of size, but also on the basis of sequence of nucleotides. Double-stranded DNA separates ("denatures") into its constituent strands when it is heated or when it is exposed to denaturing chemicals. A gradient of increasing strength of such chemicals can be produced in an electrophoretic gel so that DNA samples will travel in the direction of the electric current, separate by size, and begin to dissociate as they reach their particular critical concentration of denaturing chemical.

Every unique strand of DNA dissociates at a different concentration of denaturant. In fact, a sequence difference of only one nucleotide between two otherwise identical strands of double-stranded DNA is enough to cause the strands to dissociate at different concentrations of denaturant chemical, and to stop traveling at different locations in the gel. Using this technique, the parents' and child's DNA would be cut into fragments with restriction enzymes, dissociated into single stranded DNA, and reannealed with radioactively labeled probe DNA. The resulting heteroduplex fragments would then be separated on the basis of their DNA composition in a denaturing gradient gel. Mismatches between the sequences of probe and child's DNA and not between probe and parents' DNA would appear as different banding patterns on the gel. Again, a comparison between the banding pattern of parents' and child's DNA analyzed in this way may identify a wide range of mutations in all DNA regions.

Two-Dimensional Denaturing Gradient Gel Electrophoresis

Another approach to detecting mutations is a technique whereby sizing and denaturing gels are used to differentiate among DNA sequences common to parents' and child's DNA, polymorphisms in either parent's DNA which are transmitted to the child, and any new mutations in the child's DNA. Like the two-dimensional polyacrylamide gel procedure for protein separation described earlier, this approach compares locations of spots on a gel (in this case, DNA spots) for evidence of new mutations. In this method, various combinations of parents' and child's DNA are produced and compared on the basis of the denaturant concentration at which they dissociate. Parents' and child's DNAs are compared to each other, rather than to relatively small probes. This approach, which would allow detection of mutations in the complete genome, would detect differences ("mismatches") between a child's DNA and his or her parents' DNA. Such mismatches would represent various types of mutations in the nucleotide sequence in expressed and nonexpressed DNA regions.

DNA-RNA Heteroduplex Analysis

This technique hinges on the production of DNA bound to complementary strands of ribonucleic acid (RNA) or "DNA-RNA heteroduplexes," and on the use of specific enzymes, such as "RNaseA," that cleave the DNA at particular sequences where the DNA and RNA are not perfectly bound at every nucleotide. This is similar to the use of restriction enzymes, which cleave DNA at particular normal sequences, except that RNaseA cleaves the RNA strand in RNA/DNA hybrid molecules where there are mismatched nucleotides, indicating mutations. The resulting fragments are then separated electrophoretically to detect differences between parents' and child's DNA. The efficiency of this approach depends on the number of different mismatches that can be recognized and cleaved. This method would detect nucleotide substitutions over a large portion of the DNA.

Subtractive Hybridization

Detecting mutations would be much easier if it were possible to ignore the majority of DNA sequences that are the same in parents and child and, instead, focus only on the few sequences that may be different. Subtractive hybridization is an idea for selecting and characterizing sequences in a child's DNA that are different from either parent's DNA.

First, the double-stranded DNA of both parents is cut into fragments with restriction enzymes, dissociated into single-stranded DNA, and then mixed together with a set of single-stranded reference DNA sequences. These reference sequences represent all possible sequences of 18 nucleotides (analogous to a dictionary of 18-letter words using only 4 different letters.) Each reference sequence binds to its complementary sequence in the parental DNA and can be removed from the mixture. Any reference sequences left over, not bound to parental DNA, represent sequences not found in the parents' DNA. If the child's DNA binds with any of these left-over sequences, such hybrids would indicate that the child's DNA contains different sequences from those in the parents' DNA. These hybrids could be separated and analyzed for mutations.

This approach is the least well developed of all the ones discussed in this report, and its feasibility is unknown. If it does prove feasible, however, this approach would identify short sequences containing mutations in any part of the DNA, allowing further detailed study (e.g., by DNA sequencing) of the kinds of mutations that may occur.

Pulsed Field Gel Electrophoresis

If human DNA were short and simple, it could be cut up with restriction enzymes and separated electrophoretically into discrete bands, each representing a particular segment of the total DNA. However, human DNA is so long that when restriction-digested DNA is electrophoretically separated, the resulting fragments of the whole set of DNA form a continuous smear of bands. Even if the DNA could be cut into 100 or 200 fragments, the pieces would be too big to pass individually through the pores of a standard electrophoretic gel. A new technique, pulsed field gel electrophoresis is being developed to separate large fragments of human DNA and to examine such fragments for evidence of mutations. The procedure may detect submicroscopic chromosome mutations, including rearrangements, deletions, breaks, and transpositions. At the present time, however, the method cannot handle whole human chromosomes, though it works well with fragments of human chromosomes and with smaller whole chromosomes from lower organisms. This technique may be useful in detecting chromosome mutations that are intermediate in size between major rearrangements (observable by cytogenetic methods) and single base pair changes, potentially a large proportion of all possible mutations.

Detection and Measurement of Somatic Cell Mutations

The methods for detecting heritable mutations rely on comparing the DNA of parents with the DNA of their children to infer the kinds and rates of mutations that previously occurred in parents' reproductive cells. While this information is valuable in learning about heritable mutations, it may come months or years after the mutations have actually occurred, and this temporal separation of events makes it difficult to draw associations between mutations and their causes. Tests to detect *somatic* mutations may be useful in signaling the probability of heritable mutations. Such tests may be useful in relating exposures to specific mutagens with particular genetic events in the cell, and they may help to identify individuals and populations at high risk for mutations.[7]

It is thought that the mutation process is fundamentally similar in germinal and somatic cells. If this is true, then it may be possible to predict the risk of germinal and heritable mutations on the basis of measurements of somatic mutations, which are inferred from the frequency of mutant cells. Several investigators are currently working on methods to relate the frequency of mutant cells to the number and kinds of underlying mutations.

Several new techniques for detecting and measuring somatic mutations are described in this report. Mutant somatic cells may occur during growth and development and may appear as rarely as one in a million normal cells. Detection of somatic mutants requires methods for scanning through a million or so nonmutant cells to find a single variant cell. Two general approaches are used: 1) screening, which uses high-speed machinery to look at the total population of cells and either count or sort out the variants; and 2) selection, in which a population of cells is cultured in the laboratory under conditions that permit the growth of variant cells and that restrict the growth of the majority of cells that are nonvariants. The

[7]Even without knowing the exact relationship between rates and kinds of somatic and heritable mutations, it is reasonable to predict that people with high somatic mutation rates might be at higher risk for heritable mutations, either because of a particular environmental exposure or a genetic susceptibility to mutations.

new DNA technologies could be used to characterize the mutations in any such somatic variants found.

Different mutagens have been shown to produce distinguishable types of mutations ("mutational spectra") in human cells grown in culture. Determining mutational spectra may be useful in associating specific mutational changes in somatic cells with particular mutagenic agents to which individuals may be exposed. Such information could be useful in understanding the causes of mutational changes as well as in monitoring at-risk populations.

Data on kinds and rates of somatic mutations may provide a monitor for exposure to mutagens and carcinogens, and are relevant to the study of carcinogenesis and of aging. However, measurements of somatic mutation rates per se have little direct applicability to intergenerational (or transmitted) effects, without corresponding information on heritable mutations.

USE OF NEW TECHNOLOGIES IN RESEARCH AND PUBLIC POLICY

Feasibility and Validity Testing

The new technologies now range from ideas on paper to being in various stages of laboratory development, but none is yet ready for use in the field. A critical step before a technology is used in an investigation of a population thought to be at risk for mutations is that the technology be tested for validity and feasibility. To assure that a technology is a "valid" method, that is, that it detects the types of mutations that it theoretically is capable of detecting, and to characterize the degree to which it gives the "right" answer, it will be necessary to test a technology in a series of validation studies. A first step might be to test the technologies against pieces of DNA with known mutations, and at a later stage, in offspring of animals exposed to known mutagens. Feasibility testing will be required to make sure the technology can be efficiently scaled up for analyzing large numbers of samples.

Epidemiologic Activities

If the value of new technologies for detecting mutations is to be realized, it will be as tools for determining rates and patterns of mutations in epidemiologic studies of human beings. Once a technology has successfully passed through validation and feasibility tests, it will become a candidate for use in three major types of epidemiologic activities: surveillance, monitoring, and ad hoc studies.

Surveillance is a routine activity whose aim, in the context of this report, would be to measure the "baseline" rate of mutations in a defined population over the course of time and to facilitate rapid recognition of changes in those rates. *Monitoring* consists of observations over time in a population thought to be at increased risk of, in this case, heritable mutations, because of exposure to a known or suspected mutagen, for the purpose of helping the specific at-risk population in whatever way is possible. People living around hazardous waste disposal sites have been monitored for endpoints other than mutations, e.g., cancer and birth defects, and they would be likely candidates for mutation monitoring when technologies become available to do so efficiently. *Ad hoc studies* of a variety of designs are carried out to test hypotheses about suspected causes of mutations. Ideally, the results of ad hoc studies can be generalized to populations other than those specifically studied.

Extrapolation

Making predictions from observations of cause and effect in one system to probable effects in another is one form of extrapolation. The process involves a set of assumptions in moving from one system to another. The practical importance of extrapolation for mutagenicity is, ideally, to be able to predict mutagenic effects on human beings from the response in laboratory animals or lower test systems. The ability to extrapolate to human responses addresses one of the major goals of public health protection, the ability to identify substances harmful to human beings before anyone is exposed, thereby providing a rational basis for controlling exposure.

Extrapolation can be qualitative or quantitative. Qualitative, also called biologic, extrapolation involves predicting the *direction* of a result, for example, if a chemical causes mutations in a laboratory test, can we also expect mutations in human beings? Quantitative extrapolation involves translating a quantitative result in an animal test into a quantitative estimate of mutagenic risk in humans. Going a step further in extrapolation, can an estimate of mutagenic damage be translated into a measure of genetic disease?

A number of theoretical models for extrapolating mutagenic effects have been proposed, based on various parallel relationships. For instance, it might be true that if the relationship between somatic and heritable effects in animals were known after exposure to a specific mutagen, and if one could measure a somatic effect in human beings who had been exposed to the same substance, a heritable effect in human beings could be predicted, assuming the relationship between somatic and heritable effects is parallel in animals and human beings. Because of the paucity of data, particularly from human studies, it has been impossible to validate such an extrapolation model. The new technologies should allow a major increase in the database which, in turn, should allow researchers to more fully explore relationships among various types of test results.

Regulation

Congress has mandated public protection from mutagens in certain environmental health laws (e.g., TSCA and CERCLA), and other laws provide mandates broad enough to empower agencies to take action against mutagens in virtually any appropriate situation. Except for radiation, however, very little regulatory evaluation has taken place on the subject of heritable mutations. This is directly related to the lack of sensitive methods to detect heritable mutations in human beings, and the related difficulty in extrapolating from results in nonhuman test systems to probable human responses.

The Environmental Protection Agency (EPA) has recently issued "Guidelines for Mutagenicity Risk Assessment." EPA's approach is relatively simple and pragmatic. It requires only the types of information that can be acquired with current technologies, but allows for information from new technologies, as they become available. The guidelines require evidence of: 1) mutagenic activity from any of a variety of test systems, and 2) chemical interactions of the mutagen in the mammalian gonad. Using a "weight-of-evidence" determination, the evidence is classified as "sufficient," "suggestive," or "limited" for predicting mutagenic effects in human beings.

Although chemicals have not generally been regulated as mutagens, it is probable that exposures to mutagens have been reduced by regulations for carcinogenicity. Strong evidence supports the idea that a first stage in many cancers involves mutation in a somatic cell, and one of the most widely used screening tests for potential carcinogens, the Ames test, is actually a test of mutagenicity. The extent to which people are protected against heritable mutations if their cancer risk from a specific agent is minimized is at present unknown. The new technologies should greatly improve our ability to make that judgment.

Federal Spending for Mutation Research

OTA queried Federal research and regulatory agencies about their support of research directed at understanding human mutations. For fiscal year 1985, they reported about $14.3 million spent on development or applications of methods for detecting and/or counting human somatic or heritable mutations. An estimated $207 million was spent in the broader category of related genetic research.

OPTIONS

Research related to the new technologies described in this report has the dual aims of increasing the knowledge base in human genetics about the causes and effects of mutations, and producing information that could be used to estimate mutagenic risks for the purpose of protecting public health. The pace and direction of research toward developing these methods and the quality and efficiency of preliminary testing of methods could be influenced by congressional and executive branch actions and priorities. Integration of research in different test systems and progress in developing extrapolation models also can be influenced by actions now and in the near future.

Continued progress in the development and application of new technologies will depend not only on support of the individuals and laboratories directly involved in this research, but also on work in other areas. Although not directly addressed in this assessment, support of research in medical, human, mammalian, and molecular genetics will be essential to a full understanding of the mutation process.

The options that follow are grouped in four sections: 1) options to influence the development of new techologies, 2) options to address various aspects of feasibility and validity testing, 3) options concerning the use of new technologies in field studies, and 4) options to encourage coordination of research and validation of extrapolation models.

Development of New Technologies

The Department of Energy (DOE), several agencies of the Department of Health and Human Services, and EPA currently provide funding to researchers in independent and government laboratories for various research and development activities pertaining to human mutation research and the development of new technologies. Each agency, appropriately, proceeds down a slightly different path. OTA estimates that a total of about $14.3 million was spent on human mutation research in fiscal year 1985. Progress in this research could be speeded up by increased funding, though it is difficult to quantify an expected gain. It is clear also that less money spent on human mutation research will slow progress in laboratories already engaged in this research, and could deter other scientists from pursuing research in this field.

Option 1: Congress could assure that funding levels for human mutation research and closely related studies do not decline without the responsible agencies assessing the impact of funding cuts on research progress. This requirement could be expressed in appropriation, authorization, or oversight activities.

Of the several funding agencies, DOE has taken the lead in funding much of the research on new methods for mutation detection described in this report, and researchers at some of DOE's National Laboratories are among the leaders in the research. DOE also is the agency responsible for funding U.S. participation in the Radiation Effects Research Foundation, the joint United States-Japan body that continues to study the health of Japanese atomic bomb survivors.

Option 2: A lead agency for research related to detecting and characterizing mutations could be designated. The lead agency would be responsible for tracking the development of new technologies, facilitating the interchange of information among scientists developing the technologies and those in related fields, and encouraging and facilitating coordinated studies involving different subdisciplines. The lead agency also would keep Congress informed about activities in this area. DOE may be the logical choice for a lead agency.

The types of activities that a lead agency might engage in are described below. These activities are important whether or not a lead agency is designated, and Congress should consider directly encouraging them if there is no lead agency.

Tracking the Development of New Technologies

All the new technologies require improvements in efficiency before they become useful tools for studying human beings. As research proceeds, some techniques will develop more quickly than others, some will be dropped, and some may change in character, altering the kinds of mutations they can detect. The lead agency would be responsible for keeping track of these developments.

It would be useful if the "tracking" responsibility could lead to actions on the part of the lead agency that would promote the rapid and efficient development of the technologies. At some point, the technologies will be ready for feasibility testing and eventually, field testing. It would help researchers to know the stages of development of various methods, and it could help agencies make decisions about funding studies using certain methods. In addition, the lead agency could assist researchers by anticipating needs that will be common to all research programs and by encouraging efficient use of resources.

Facilitating Information Exchange Among Researchers

In 1984, DOE organized and funded a meeting that brought together for the first time many of the researchers involved in laboratory-based mutation research. This meeting is acknowledged among those who attended as a milestone for information exchange and the generation of new ideas concerning detection of heritable mutations. In fact, the ideas for some of the new methods described in this report were born at that meeting. There is a continuing need for this type of information exchange.

Keeping Congress Informed

Congress has already directed regulatory agencies to reduce exposures to environmental agents that may cause mutations. As part of its oversight of both regulatory and research activities, Congress could benefit from up-to-date information about the development of various technologies. The lead agency could report in a specified manner, e.g., a brief annual report describing the level of current research, its goals, results of completed or ongoing work, and expected near-term developments. This information could also be the basis for informing the public about mutation research. This activity will continue to be valuable to Congress through later phases of development and application of new technologies.

Feasibility and Validity Testing

After technologies pass through a phase of development to improve their efficiency and to work out technical details, a period of feasibility and validity testing will be necessary before a technology can rationally be used as a tool in a large-scale study of heritable mutations in human beings. There are some ways to make validity testing an efficient process. As an example, a "DNA library," a collection of known DNA sequences, particularly sequences carrying known mutations, could be established and maintained by one laboratory, which could make DNA sequences available to all researchers developing mutation detection technologies. This material could be used to determine whether a particular technology detects those mutations that it is designed to detect, analogous to testing chemical procedures and equipment against known chemical "standards." At a more advanced stage, new technologies might also be tested in animal experiments with a selection of mutagens known to cause specific types of mutations. If a lead agency is designated, some of these options would logically be among that agency's responsibilities. If there is no lead agency, these functions could still be encouraged by Congress through oversight activities.

Banking Biological Samples

Biological samples, especially blood samples, are often collected during the course of medical examinations for people thought to be exposed to environmental or occupational agents. Implementation of a plan to bank those samples would facilitate human mutation studies and other related research.

Examining stored samples is not nearly so disruptive as collecting samples from a population. The very act of specifically collecting samples for a study of mutations would raise expectations that definitive information about risks would be forthcoming. Examining stored samples would avoid that human cost and could, at the same time, provide a realistic test of new technologies before they are applied to people who are anxious about the effects of environmental exposures on their genes.

Although stored samples offer many advantages, preparing human samples for storage and maintaining the stored samples is a significant task. Currently the high cost of storing samples inhibits establishment of sample banks. There has been little research directly aimed at improved methods for storing biological samples.

Option 3: Congress could encourage banking of biological material obtained from at-risk individuals and their spouses and offspring, with the objective of studying somatic and heritable mutations from these stored samples before technologies are used directly in a study of an exposed population. A centralized samples bank available to all researchers would be the most efficient means of establishing a repository of sufficient size and scope for validity testing.

Option 4: Congress could encourage the appropriate agencies (through the lead agency, if one is designated) to investigate the potential for improving technologies for storing biological samples. Where possible, funding should be encouraged for such improvements. This option is relevant to a broad spectrum of human health research, and collaboration among disciplines should be encouraged for determining the specific storage needs of different lines of research.

Field Studies

Sometime in the next 5 to 10 years, it is likely that one or more of the technologies for studying human mutations will be brought to the point of readiness for epidemiologic studies of human beings thought to be at risk for mutations. Such a study could be initiated by an independent scientist or a group of scientists involved in the development of a new technology who have identified an at-risk population; a new technology or technologies could be used in an environmental health investigation triggered by the discovery of a potentially at-risk population; or Congress could mandate a study of mutations in a specific population. Depending on the way a study is initiated, different agencies will be drawn in, and different funding mechanisms will come into play.

Should a study be initiated by one or a group of investigators who submit a grant proposal for funding, the current system for review of research awards is probably appropriate for making scientific judgments about whether a method is ready for large-scale testing. From that point on, however, there are significant differences between a proposal to use a new technology in a human population and most other proposals. The major differences are size and money. Application of any promising technology will require that state-of-the-art methods in molecular biology be expanded from small-scale laboratory use to large-scale examination of blood samples collected from hundreds or thousands of people.

A study such as that described above would require a significant commitment of funds over a period of years, and could account for a large percentage of a granting agency's funds. The amount of money necessary can only be estimated, but it could amount to tens of millions of dollars, not a large amount in government spending, but large in comparison to most biomedical research grants, which usually range from less than $100,000 to about $500,000 annually. The necessary commitment of funds from any single agency, considering current spending for this type of research, is likely to be impossible, no matter how worthwhile the proposed study.

Option 5: Congress could consider providing specific add-on funding to finance an investigator-initiated realistic test of a promising method to study human heritable mutations.

A study mandated by Congress or an agency-initiated study using new technologies will not necessarily undergo the rigorous review and criticism that a proposal submitted to a granting agency would. Congress already has some experience in mandating epidemiologic studies. Concerns about cancer and birth defects convinced Congress to mandate studies of military veterans who were exposed to Agent Orange or who participated in atomic bomb tests. In addition, Congress has vested the Public Health Service with the authority to carry out a wide range of health investigations of exposures from toxic waste dumps through "Superfund." The role of the mandated "Agency for Toxic Substances and Disease Registry" (ATSDR) is to "effectuate and implement the health related authorities" of Superfund. ATSDR, located within the Centers for Disease Control, is a logical place for new technologies to be used as exposed populations are identified through other provisions of Superfund.

An unusual array of health problems or an apparent excess of disease among people living around toxic waste sites could trigger a study by ATSDR. Such problems could also prompt local residents to press Congress for studies to determine whether, among other things, the disorders had been caused by mutations, and whether these mutations could have resulted from chemicals in the waste site.

Some scientific societies, such as the American Society of Human Genetics, could be asked to participate in a review of the feasibility of detecting possible health effects from environmental exposures. If the study is determined to be feasible, Congress and the public can be reassured that the study's findings are likely to be useful in decisionmaking. Alternatively, a decision that no study is currently feasible would underline the importance of developing and testing new methods. Reviewing the feasibility of a study may be perceived as delaying an investigation unnecessarily. However, performing a study that has little or no likelihood of answering questions about exposure and mutations has marginal value at best and would not serve the people who requested it.

Option 6: If Congress mandates a study of heritable mutations in an at-risk population using the new technologies, the mandate should include a feasibility assessment by a panel of experts before the study begins.

Option 7: Congress could require agencies to plan for a rigorous outside review by a panel of experts of any agency-initiated study using a new technology, before such a study can proceed.

Integration of Animal and Human Studies

Much of our current ability to estimate the effects of various external agents on human beings is derived from animal studies. In the future, animals will continue to be used to test for mutagenicity and, ideally, to predict effects in human beings. Right now, the available methods and body of data provide an inadequate basis for making predictions from animal results to human effects. The new technologies described in this report for detecting heritable and somatic mutations can be applied in both animal systems and in human beings. This information and information from currently available animal systems could be used in an integrated fashion to study relationships between somatic, germinal, and heritable mutations, and to pursue the development of extrapolation models for predicting effects in human beings.

The kind of integrated research necessary will not occur spontaneously if the current pattern holds. There are few researchers engaged in studies that integrate comparable information from different systems for the purpose of elucidating relationships among such systems. Improvements in extrapolation from animal to human data and from somatic to heritable mutations await efforts to encourage and coordinate the appropriate research. Single laboratories or centers are unlikely to be able to perform all the different tests necessary to develop and test extrapolation models, making coordination among laboratories essential.

The National Toxicology Program (NTP) is the center of Federal mutation research using experimental animals, funding research grants and contracts nationally and internationally. The National Center for Toxicological Research, a laboratory of the Food and Drug Administration that is part of NTP, has facilities and experience to carry out large-scale animal tests. EPA also has a genetic toxicology program, and it is actively pursuing development and application of methods for studying human somatic mutations. DOE's National Laboratories have directed or carried out the majority of large-scale animal studies of mutation rates and mechanisms, and this experience, as well as the advanced technology that these laboratories have developed for sorting and studying chromosomes and cells, will be useful in improving methods for extrapolation. These three agencies could encourage the development of methods for linking information from animal studies to somatic and heritable mutation risks in humans.

Option 8: Congress could encourage studies of somatic, germinal, and heritable mutations in experimental animals using both currently available and new technologies. Further, research funding agencies should encourage animal studies directed at identifying the mechanisms of mutagenesis and elucidating relationships between mutagenic potencies in animals and in human beings.

Chapter 2

Genetic Inheritance and Mutations

Contents

FIGURES

Chapter 2

Genetic Inheritance and Mutations

"Like begets like" is an expression that describes repeated observations in families. Except in rare instances, most children are normal and generally resemble their biological parents, brothers, and sisters. In these cases, the genetic information has been passed intact from the parents to their children.

This report is about the instances when something goes wrong, when a genetic mistake is made; a mutation in the DNA of the parents' reproductive cells, both of whom are physiologically normal, is passed to a child who may or may not appear normal, depending on the nature and severity of the mutation. All types of mutations, from the benign to the severe, are considered in this report. As an introduction to human genetics for nonspecialists, this chapter summarizes some basic information about normal functions of DNA, the various kinds of mutations that can occur, and health effects of mutations. More complete information can be found in a general reference such as Vogel and Motulsky (165).

FUNCTION AND ORGANIZATION OF DNA

Deoxyribonucleic acid, DNA, is the carrier and transmitter of genetic information. Its functions require that it is relatively stable and that it is able to produce identical copies of itself. Certain agents, such as some forms of radiation, viruses, and certain chemicals, alter DNA's ability to maintain these characteristics. Mutations, changes in the composition of DNA, can occur as a result of these agents acting on DNA or on the systems in place to repair DNA damage.

Human DNA can be thought of as an immense encyclopedia of genetic information; each person's DNA contains a unique compilation of roughly 3 billion nucleotides in each of its two chains, the exact sequence shared by no other person except an identical twin. The genetic information encoded in DNA is contained in the nucleus of cells, and is packaged in units, or *chromosomes*, that consist of long twisted double strands of DNA surrounded by a complex of proteins. Each chromosome contains thousands of *genes*, the functional units of DNA, which are "read," or transcribed, by the cell so that genetic information can be used to make proteins or to regulate cell functions. It is thought that only a small proportion, approximately 1 to 10 percent, of human DNA is translated into specific proteins. Functions of the nontranslated majority of DNA are largely unknown. There are 23 pairs of chromosomes in all nucleated human cells except human reproductive cells (germ cells). The latter contain one of each pair of chromosomes, since two germ cells (one egg and one sperm) fuse at conception and re-create a full set of DNA in the offspring.

DNA is composed of two chains of nucleotides bound together in a double helical structure (fig. 2). The backbones of the two chains, formed by sugar (deoxyribose) and phosphate molecules, are held to each other by hydrogen-bonded *nitrogenous bases*: adenine, thymine, guanine, and cytosine, abbreviated A, T, G, and C, respectively. Each of the chains' units, or *nucleotides*, consist of a sugar, phosphate, and nitrogenous base. The linear sequence of nucleotides repeated in various combinations thousands of times along the DNA determines the particular genetic instructions for the production of proteins and for the regulation of cell functions.

A change in even a single nucleotide, e.g., substituting G for T, among the 3 billion nucleotides in the DNA sequence, constitutes one type of mutation. Depending on where it occurs, such a mutation could be sufficient to cause significant pathological changes and dysfunction, or it could cause no impairment at all. In thalassemia, for example, a genetic disorder of hemoglobin synthesis, approximately 40 different "spelling er-

Figure 2.—The Structure of DNA

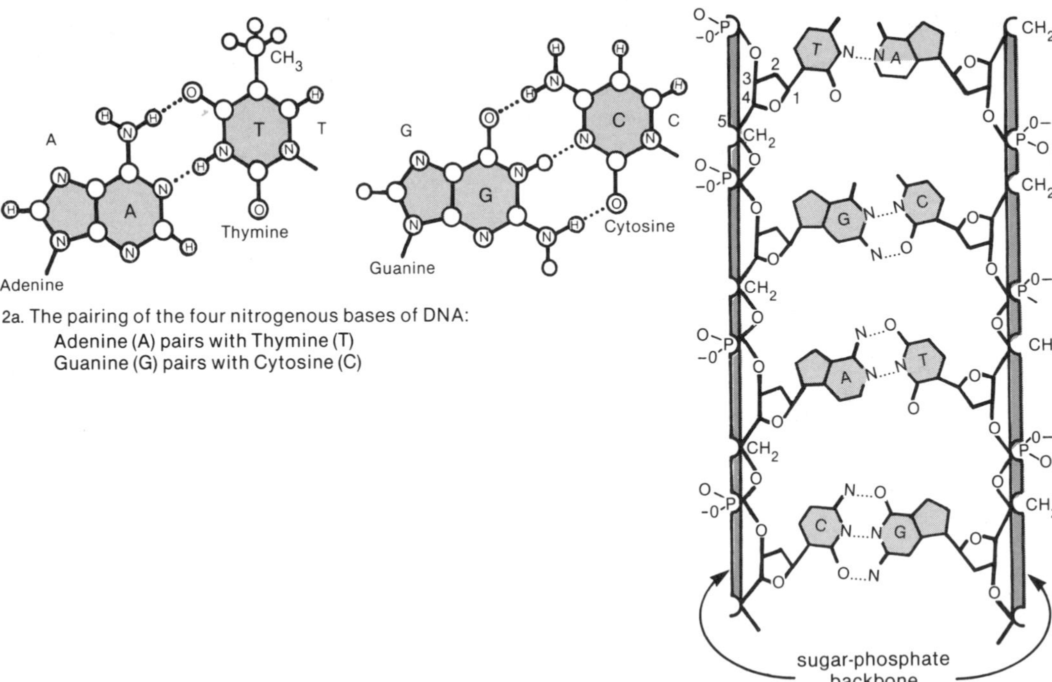

2a. The pairing of the four nitrogenous bases of DNA:
Adenine (A) pairs with Thymine (T)
Guanine (G) pairs with Cytosine (C)

2b. The four bases form the four letters in the alphabet of the genetic code. The *sequence* of the bases along the sugar-phosphate backbone encodes the genetic information.

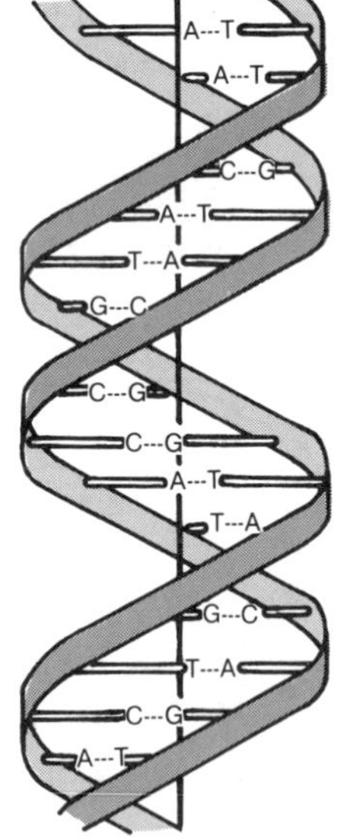

A schematic diagram of the DNA double helix.

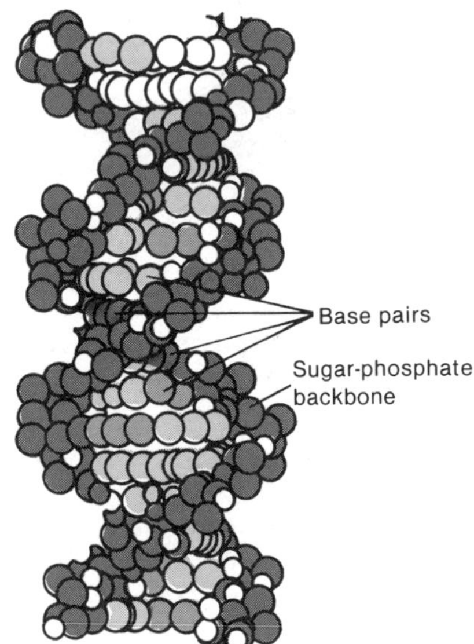

A three-dimensional representation of the DNA double helix.

2c. The DNA molecule is a double helix composed of two chains. The sugar-phosphate backbones twist around the outside, with the paired bases on the inside serving to hold the chains together.

SOURCE: Office of Technology Assessment.

rors," or (single or multiple) nucleotide changes, have been identified in a particular gene that specifies the production of globin, an essential constituent of hemoglobin. In complex diseases such as heart disease, susceptibility is thought to be influenced by the interaction of environmental factors such as diet, and genetic factors such as mutations in one or several genes.

During DNA replication, each chain is used as a template to synthesize copies of the original DNA (fig. 3). New mutations are transmissible to daughter cells. To replicate, the two chains of the DNA double helix separate between nucleotides, and each is copied by a series of enzymes that insert a complementary nitrogenous base opposite each base in the original strand, creating two identical copies of the original one. As a result of specific pairing between nitrogenous bases, each chain builds its complementary strand using the single chain as a guide.[1]

Protein Synthesis

The normal functioning of the human body—including the digestion of food; the production of muscle, hair, bone, and skin; and the functioning of the brain and nervous system—is dependent on a well-ordered series of chemical reactions. Proteins, especially the enzymes, promote the chemical reactions on which these processes are based. Each enzyme has a specific function; it recognizes a chemical, attaches to it, reacts with it, alters it, and leaves, ready to promote the same reaction again with another chemical.

To synthesize a protein, the genetic information contained in one or more genes along the DNA is transcribed to small pieces of ribonucleic acid (RNA), which are faithful replicas of one strand of DNA (see fig. 4). Each strand of RNA moves out of the nucleus to the cytoplasm of the cell, where it serves as a template for protein synthesis. Proteins are composed of hundreds of linked subunits called *amino acids*. At the ribosomes, RNA is used to gather and link amino acids in a specific sequence and length according to the design specified in the DNA to form the different proteins.

[1]The molecular structure of the nitrogenous bases in DNA requires that A pairs only with T (or with U in RNA) and G pairs only with C.

Figure 3.—Replication of DNA

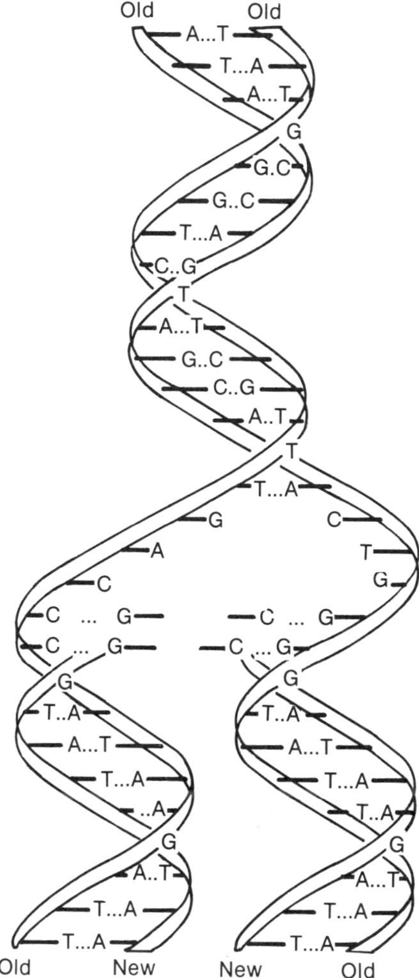

When DNA replicates, the original strands unwind and serve as templates for the building of new complementary strands. The daughter molecules are exact copies of the parent, with each having one of the parent strands.

SOURCE: Office of Technology Assessment.

Genetic information in DNA is organized in a triplet code, or sequence of three nucleotides. Each triplet specifies a complementary *codon* in "messenger RNA" (mRNA) which, in turn, specifies a particular amino acid. The sequence of such triplets in a gene ultimately determines the amino acid sequence of the corresponding protein. Codons exist for each of the 20 amino acids that make up the myriad of different proteins in the body. Additional codons exist to signal "start" and

Figure 4.—Diagrammatic Representation of Protein Synthesis According to the Genetic Instructions in DNA

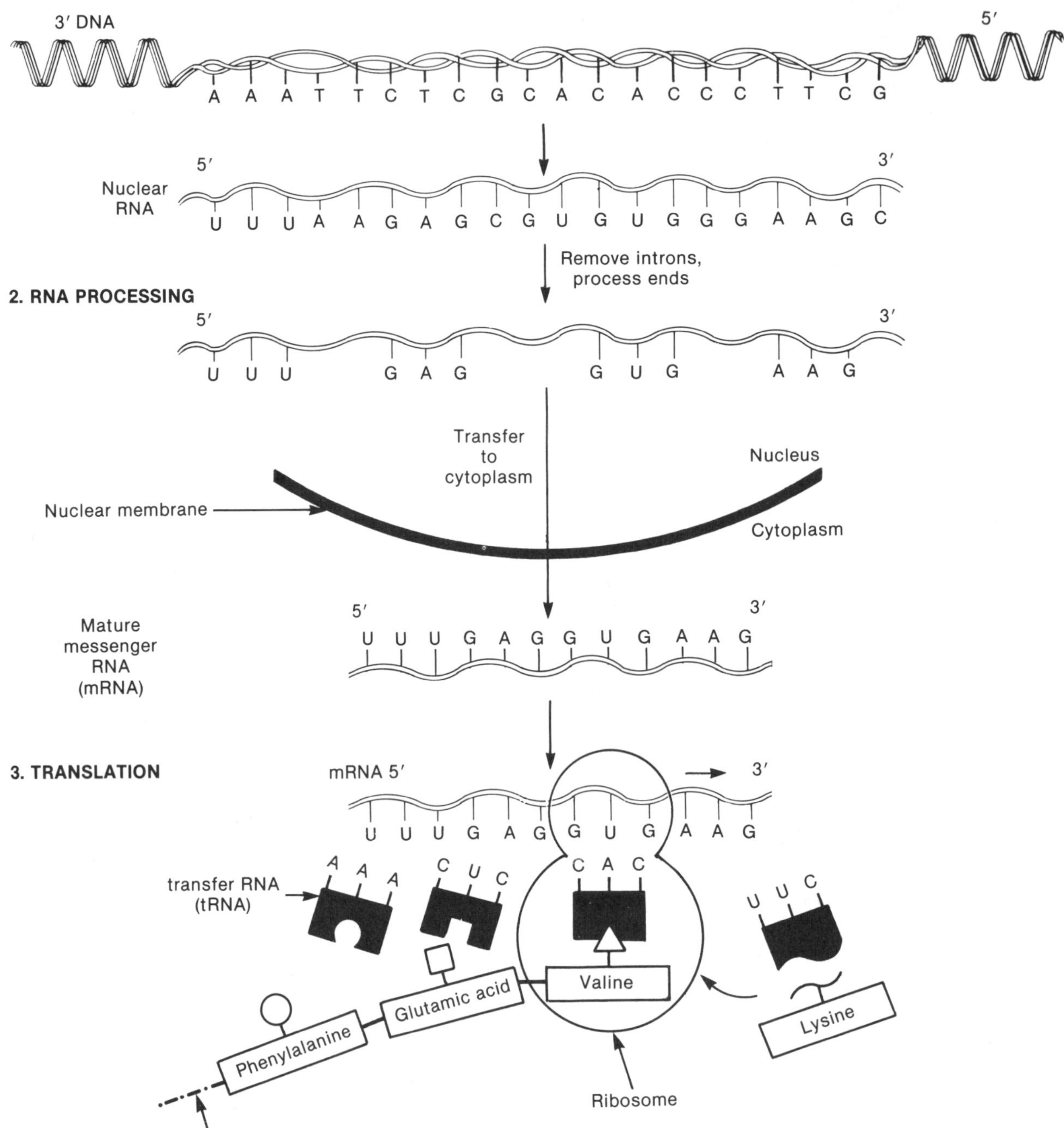

SOURCE: Adapted from A.E.H. Emery, *An Introduction to Recombinant DNA* (Chichester: John Wiley & Sons, 1984).

Figure 5.—The Genetic Code

Codon	Amino Acid	Codon	Amino Acid	Codon	Amino Acid	Codon	Amino Acid
UUU	Phenylalanine	UCU	Serine	UAU	Tyrosine	UGU	Cysteine
UUC	Phenylalanine	UCC	Serine	UAC	Tyrosine	UGC	Cysteine
UUA	Leucine	UCA	Serine	UAA	stop	UGA	stop
UUG	Leucine	UCG	Serine	UAG	stop	UGG	Tryptophan
CUU	Leucine	CCU	Proline	CAU	Histidine	CGU	Arginine
CUC	Leucine	CCC	Proline	CAC	Histidine	CGC	Arginine
CUA	Leucine	CCA	Proline	CAA	Glutamine	CGA	Arginine
CUG	Leucine	CCG	Proline	CAG	Glutamine	CGG	Arginine
AUU	Isoleucine	ACU	Threonine	AAU	Asparagine	AGU	Serine
AUC	Isoleucine	ACC	Threonine	AAC	Asparagine	AGC	Serine
AUA	Isoleucine	ACA	Threonine	AAA	Lysine	AGA	Arginine
AUG	Methionine (start)	ACG	Threonine	AAG	Lysine	AGG	Arginine
GUU	Valine	GCU	Alanine	GAU	Aspartic acid	GGU	Glycine
GUC	Valine	GCC	Alanine	GAC	Aspartic acid	GGC	Glycine
GUA	Valine	GCA	Alanine	GAA	Glutamic acid	GGA	Glycine
GUG	Valine	GCG	Alanine	GAG	Glutamic acid	GGG	Glycine

Each codon, or triplet of nucleotides in RNA, codes for an amino acid (AA). Twenty different amino acids are produced from a total of 64 different RNA codons, but some amino acids are specified by more than one codon (e.g., phenylalanine is specified by UUU and by UUC). In addition, one codon (AUG) specifies the "start" of a protein, and 3 codons (UAA, UAG, and UGA) specify termination of a protein. Mutations in the nucleotide sequence can change the resulting protein structure if the mutation alters the amino acid specified by a triplet codon or if it alters the reading frame by deleting or adding a nucleotide.

U - uracil (thymine) A - adenine
C - cytosine G - guanine

SOURCE: Office of Technology Assessment and National Institute of General Medical Sciences.

"stop" in the construction of a polypeptide (protein) chain. The triplet code provides more information, however, than is needed for 20 amino acids; all except 2 of the amino acids are specified by more than one codon (see fig. 5).

Inheritance of Genetic Traits

A child's entire genetic endowment comes from the DNA in a single sperm and egg that fuse at conception. The fertilized egg then divides, becoming a multicellular embryo, and the cells differentiate into specialized tissues and organs during embryonic development. In terms of genetic material, there are two general types of cells: somatic cells and germinal cells. Those that have a full set of DNA at some time in their development and do not participate in the transmission of genetic material to future generations are *somatic cells*, which include all cells in the body except the reproductive cells. *Germ cells* include the gametes (egg and sperm, each of which contains half of the total set of DNA) and the cell types from which the gametes arise.

A normal, full set of human DNA in somatic cells consists of 46 chromosomes, or 23 pairs of homologous chromosomes: 22 pairs of "autosomes" and 1 pair of sex-determining chromosomes. Females have 22 pairs of autosomes and two X chromosomes, while males have 22 pairs of autosomes and an X and a Y chromosome. Reproductive cells, with half the normal set of DNA, are produced by reduction division, or meiosis, from sex-cell progenitor cells in the female ovary or male testis. These progenitor cells are set aside early in fetal development and are destined to be the only source of germ cells. Each ovum or sperm has 23 single, unpaired chromosomes: ova have 22 autosomes and one X chromosome; sperm have 22 autosomes and one X or one Y chromosome.

Most genes are present in two copies, one on each member of a homologous pair of chromosomes. For instance, if gene A is found on chromosome 1 it will be present at the same location or *locus* on both copies of chromosome 1. The exact nature of the DNA sequence of gene A may differ between the two chromosomes, and the word *allele*, classified as either *normal* or *mutant*, is used to refer to different forms of the same gene. Many different mutant alleles are possible, each one defined by a particular change in the DNA sequence. A mutation that is not expressed when a normal gene is present at the same locus on the sister chromosome is called *recessive* and a mutation that is expressed even in the presence of a normal gene is called *dominant*.

A Normal Karyotype of 46 Human Chromosomes

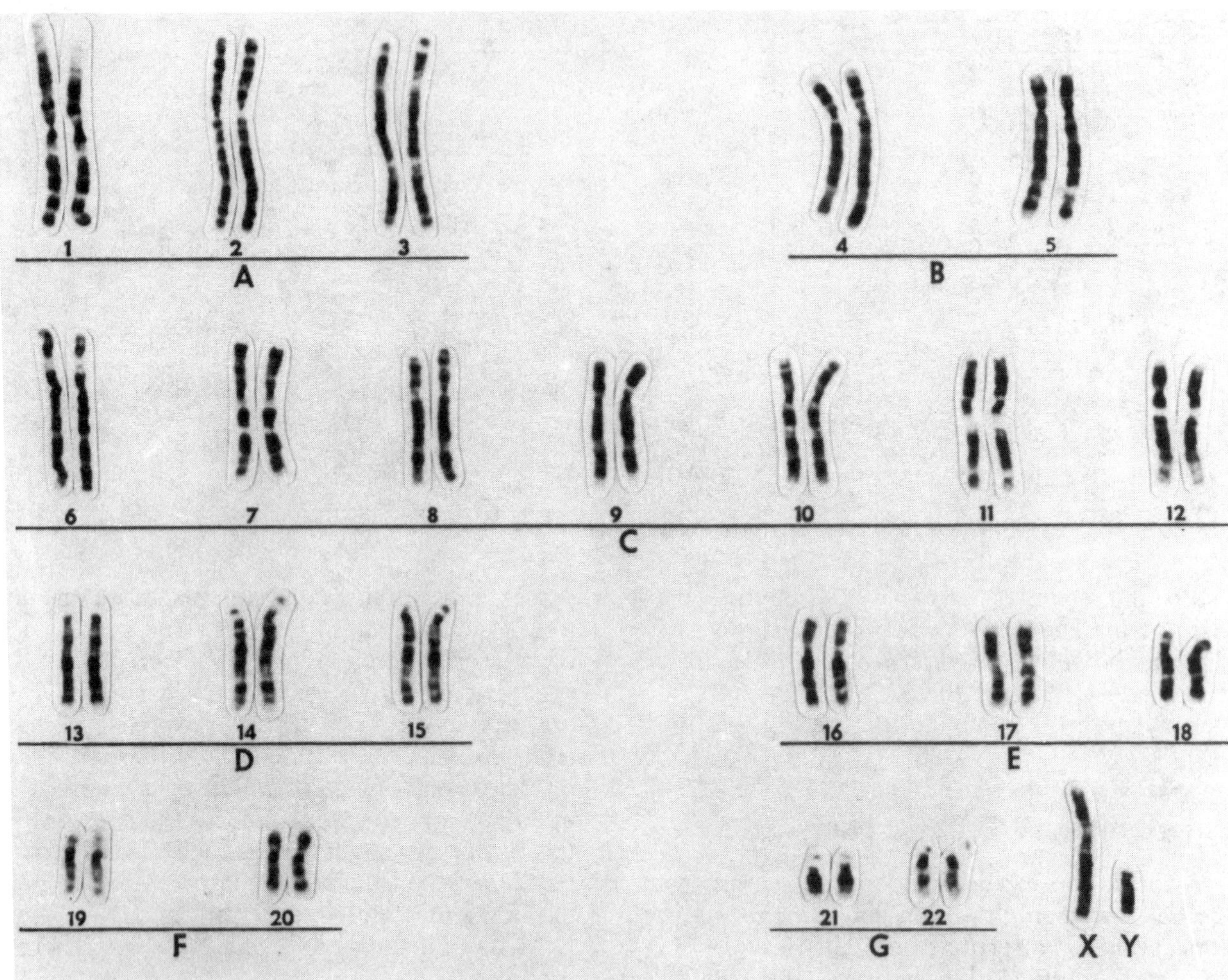

Photo credit: Gail Stetten, Johns Hopkins Hospital

Karotypes typically show all the chromosomes of a single cell, and are used mainly to identify abnormalities in the number or structure of chromosomes. Chromosomes can be isolated from any nucleated cell type in the body, but are most often isolated from white blood cells derived from a sample of venous blood or from amniotic fluid cells obtained at amniocentesis. The cells are stimulated to grow and divide in the laboratory, yielding a large number of cells which contain a sufficient amount of DNA to isolate and examine. The cells are then blocked in their dividing phase by the addition of a chemical, such as colcemid, to their growth media. After harvesting the cells, they are treated with a hypotonic solution to cause them to swell, and then with a chemical fixative to maintain this fragile, swollen state. When small amounts of the cell suspension are carefully dropped onto microscope slides, the cells break open, spreading out their chromosomes. To visualize the chromosomes and to distinguish the different homologous pairs, the chromosomes are stained to produce a characteristic banding pattern for each of the 23 pairs of homologous chromosomes. The chromosomes can then be photographed under the microscope, cut out from the final print, and numbered and arranged in order of decreasing size as shown in this karyotype.

KINDS OF MUTATIONS

Mutations are changes in the composition of DNA. They may or may not be manifested in outward appearances. Some mutations, depending on where they occur in the DNA, have no apparent effects at all, and some cause changes that are detectable, but are without obvious impairment. Still other mutations lead to profound effects on human health and behavior: stillbirths, neonatal and early childhood deaths, various diseases, severe physical impairments, and mental deficiencies.

If mutations occur in genes that determine the structure or regulation of essential proteins, the results are often detrimental to health. Since most DNA in human cells is not expressed or used in any known way, mutations in these regions may or may not be clinically apparent, although it is possible that subtle physiological variations may occur. Interactions among several such mutations may result in unusual responses to environmental stimuli or they may influence susceptibility to various chronic diseases.

Mutations can be generally divided according to size into two groups: gene mutations and chromosome mutations. Chromosome mutations include numerical and structural abnormalities.

Gene Mutations

Gene mutations occur within or across a single gene, resulting from the substitution of one nucleotide for another or from the rearrangements (e.g., deletions) within the gene, or they may duplicate or delete the entire gene. The mutation responsible for sickle cell anemia, for example, is a point mutation: a single nucleotide in the gene coding for globin, a constituent of hemoglobin, is substituted for another nucleotide, resulting in the substitution of one amino acid (glutamic acid) for another (valine) at a certain position in the globin chain. One of the mutations responsible for a form of alpha-thalassemia, another form of anemia, involves the deletion of an entire alpha-globin gene, resulting in the absence of alpha-globin synthesis.

Since proteins are produced from the instructions in genes, a mutation in a gene that codes for a specific protein may affect the structure, regulation, or synthesis of the protein. A mutation that results in a different codon may or may not change the resulting protein, since different codons can code for the same amino acid. Alternatively, such a mutation may result in the substitution of one amino acid for another, and possibly change the charge distribution and structure of the protein. Another mutation in the same gene may result in gross reduction or complete loss of activity of the resulting protein. Detailed analysis of the mutations underlying thalassemia disease has shown that mutations occurring in the bordering areas, not within the translated parts of the gene, can also impair gene function (171).

Chromosome Mutations

Major changes affecting more than one gene are called chromosome mutations. These involve the loss, addition, or displacement of major parts of a chromosome or chromosomes. They are often large enough to be visible under the light microscope as a change in the shape, size, or staining pattern ("banding") of a chromosome. Major chromosome abnormalities are often lethal or result in conditions associated with reduced lifespan and infertility.

Structural Abnormalities

Structural mutations in the chromosomes change the arrangement of sets of genes along the chromosome. Sections of one chromosome, including many genes, may break and reattach to another chromosome (e.g., in a balanced translocation, shown in fig. 6) or another location on the same chromosome. Alternatively, sections of chromosomes can be deleted, inserted, duplicated, inverted.

Numerical Abnormalities

Numerical chromosomal mutations increase or decrease the number of whole chromosomes,

Figure 6.—Chromosomal Translocation

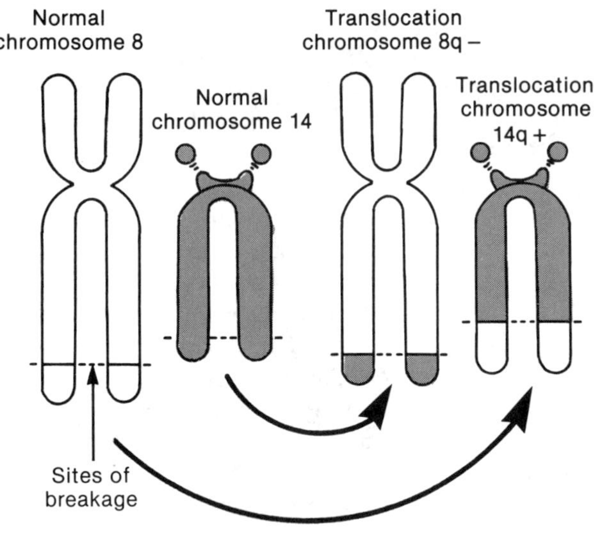

SOURCE: Office of Technology Assessment.

without changing the structure of the individual chromosomes. Errors in the production of the germ cells can lead to abnormal numbers of chromosomes in the offspring. For example, Down syndrome can be caused by the presence of an extra chromosome (Trisomy 21); three copies of chromosome 21 are present while all other chromosomes are present in normal pairs, giving a total of 47 chromosomes in each cell. A similar, but opposite, error in production of the germ cells results in the lack of one chromosome. Turner syndrome (Monosomy X), in which affected females have 44 autosomes and only one sex chromosome, an X chromosome, results in abnormal development of the ovaries and in sterility. The effects of these mutations may result not from alteration in the nature of the gene products but from the abnormal amount of gene products present.

HEALTH EFFECTS OF MUTATIONS

In Western countries, genetic diseases are more visible now than in the earlier part of this century. Advances in medical care, public health measures, and living conditions have contributed to a gradual decline in the contribution of environmental factors to certain types of diseases, particularly, infectious diseases and nutritional deficiencies (32). Genetic disorders as a group now represent a significant fraction of chronic diseases and mortality in infancy and childhood, although each disorder is individually rare. Such disorders generally impose heavy burdens, expressed in mortality, morbidity, infertility, and physical and mental handicap. Approximately 90 percent of them become clinically apparent before puberty.

The incidence of genetic disease is not known precisely, but it is estimated that genetic disorders are manifested at birth in approximately 1 percent of liveborn infants, and are thought to account for about 7 percent of stillbirths and neonatal deaths and for about 8.5 percent of childhood deaths. Nearly 10 percent of admissions to pediatric hospitals in North America are reported for genetic causes. The majority of these genetic diseases lack effective treatment. A small, but increasing number are avoidable through prenatal diagnosis and selective abortion (21,32,40).

In most of these cases, the particular genetic disorder is passed along in families, or the asymptomatic carrier state for the disease is passed along for many generations, before it may be expressed in a child with the disorder. In some of these cases, however, the disorder may suddenly appear in a child when no relatives have had the disorder. Not uncommonly, such an event can be explained by mistaken parentage or by one of the parents showing mild or almost unnoticeable signs of the disorder, so that this parent has a disease-causing gene that can be passed on to his or her children. However, the sudden, unexpected appearance of a new condition in a family can also result from a new mutation in the reproductive cells of the parents. Examples of conditions that result from new mutations in a large percentage of cases include Duchenne muscular dystrophy, osteogenesis imperfecta (a condition resulting in brittle bones), and achondroplasia (dwarfism).

The exact causes of new heritable mutations are unknown. It is thought that the risk of some new heritable mutations increases with increasing age of the father. Certain rare disorders due to mutations occur at as much as four times the average rate when fathers are 40 years of age or older at the time of conception of their children. Such disorders include achondroplasia, Apert's syndrome (a disorder with skull malformations and fusion of bones in the hands), and Marfan's syndrome (a complex syndrome including increased height). Advanced *maternal* age is a risk factor for some newly arising chromosome abnormalities, including Down syndrome.

Very small deletions of a particular chromosome have been found in children having certain cancers such as Wilm's tumor (chromosome number 11) and retinoblastoma (chromosome number 13). Some chromosome mutations are compatible with normal life, and others may be associated with increased morbidity and reproductive problems. For example, when a section of a chromosome breaks off and reattaches to another chromosome, both copies of all of the genes may still be present, but they are located abnormally. If no genetic material has been lost in the move, this type of rearrangement is called a balanced translocation. However, a gene that was previously "turned off" may, in its new location, be suddenly "turned on"; this explanation has been invoked to explain how cancer genes can be activated. Furthermore, in the production of germ cells, an abnormal complement of DNA could result, so that a parent with a balanced translocation may have offspring with unbalanced chromosomes, an outcome that usually has deleterious consequences for the child.

Early Mortality and Morbidity

Numerical and structural chromosome abnormalities have been detected in as many as 50 to 60 percent of spontaneous abortions, 5 to 6 percent of perinatal deaths, and 0.6 percent of liveborn infants (43). Fetuses that survive to term with such abnormalities are apparently only a small percentage of the number of fetuses with chromosome abnormalities that have been conceived. Abnormalities such as trisomy 13 or trisomy 18 are among the more common disorders found in the liveborn group, whereas many numerical and structural abnormalties, most of which are not found in surviving infants, have been found in spontaneous abortuses.

Some of the numerical chromosome abnormalities, such as trisomy 13 or trisomy 18, can be compatible with a full-term pregnancy but may cause severe handicap in infancy and childhood and may be associated with a reduced lifespan. The health effects of structural chromosome aberrations (involving only parts of chromosomes) depend on the size and location of the aberration. However, features shared by many of these conditions include low birthweight; mental retardation; and abnormalities of the face, hands, feet, legs, arms, and internal organs, often neccessitating medical care throughout an individual's life.

Diseases During Adult Life

Some genetic disorders, although present in an individual's DNA since his conception, do not manifest themselves until adult life. Huntington disease, a degenerative neurological disorder characterized by irregular movements of the limbs and facial muscles, mental deterioration, and death usually within 20 years of its first physiological appearance, is an example of a genetic disorder with adult onset.

Many disorders that develop during adulthood and that tend to cluster in families do not have classical patterns of genetic inheritance. Several genes may be involved in the development of these "multifactorial diseases," and environmental factors may also influence the manifestation and severity of these disorders. Susceptibilities to conditions such as diabetes mellitus, hypertension, ischemic heart disease, peptic ulcer, schizophrenia, bipolar affective (manic depressive) disorder, and cancer are thought to have genetic as well as environmental components.

Effects of Somatic Mutations

Mutations that occur in somatic cells may affect the individual, but they are not passed on to future generations. Somatic mutations may play a role in the development of some malignant tumors by removing normal inhibition to cell growth and regulation, producing cells with a selective growth advantage. A wide range of experiments in animals and observations in humans indicate that somatic mutation may be one step in the development of certain types of cancer, including leukemia and cancers of the breast and thyroid. Mutations that cause deficiencies in the normal DNA repair system, such as in the genetic disease xeroderma pigmentosum, can also lead to cancer following exposure to a mutagenic agent.

PERSISTENCE OF NEW MUTATIONS IN THE POPULATION

The "spontaneous mutation rate" represents the sum of "natural" error arising during the course of life and reproduction, and external influences, both natural and manmade. The mutation rates that have prevailed throughout recent human history may be in a state of relative equilibrium, although accurate information is lacking. Suppose, though, that mutagens are suddenly introduced heavily into the environment, and that there is an increase in mutation rates. It is impossible to accurately predict the number of people born with genetic diseases in future generations that will result from that increase, but some predictions can be made based on what is known about the effects of different types of mutations on subsequent generations. Two cases are considered below: 1) the effect of a doubling of the spontaneous mutation rate in one generation, and 2) the effect of a permanent doubling of the spontaneous mutation rate.

Assume that a population was exposed to a "pulse" of a mutagenic agent, such as radiation from an atomic bomb or from chemical agents in a toxic spill, and that the exposure caused a doubling of the heritable mutation rate. Overall, the number of extra cases of disease associated with mutation in the first generation after the pulse (the generation in which the effect would be greatest), will be comparatively small. One estimate (from a 1977 United Nations Scientific Committee on the Effects of Atomic Radiation report) is that 10.5 out of every 100 liveborn infants has some kind of genetic disorder (88). Almost all of those cases are from mutations already present in the gene pool, and not from new mutations. The National Research Council (NRC) estimates that one mutational pulse would add an extra 0.66 cases, so instead of 10.5 per 100, 11.16 cases would occur. The effects on subsequent generations would be smaller, but even after hundreds of generations, there might be some small increment of the mutational load as a result of the pulse. To understand this phenomenon, it is useful to consider different kinds of mutations separately.

Chromosome mutations would be the shortest lived in the population, because in most cases they result in sterility, so they would not be passed on to the next generation. They would probably be eliminated completely after about 1.25 generations.

Dominant and X-linked (sex-linked) mutations are the next shortest lived. Many of these, because they often cause such severe disease, interfere with a normal lifespan and reproduction. After about four or five generations, on average, they would no longer be transmitted to future generations.

New recessive mutations will probably have the greatest chance of being maintained in the population. Virtually none would be eliminated in the first generation, because a single individual will have only one copy of the gene, and two would be needed to cause overt disease. Recessive mutations may take generations to surface as over disease. There is much less certainty about the ultimate fate of recessive mutations than about dominant, X-linked, or chromosome mutations.

Many "common" or multifactorial diseases are partially influenced by genetic traits carried in families or arising anew. These include congenital malformations, and such common diseases as heart disease, diabetes mellitus, bipolar affective disorder, immunological disorders, and cancer. Since it is difficult to estimate the proportion of disease that is currently influenced by mutations, it is also difficult to estimate how long such mutations induced from a pulse of a mutagenic agent would persist in the population. However, the NRC report (88) estimates that these mutations would persist for an average of 10 generations.

These estimates suggest a surprisingly small increase over background rates of genetic disease in the first few generations after intense exposure to mutagenic agents, which would decline slowly over many generations. The NRC report (88) suggests that about half of the total effect would occur in the first six generations. The most severe mutations would be more easily observable as dominant genetic diseases. These would be eliminated within one or two generations since they interfere with survival and fertility. Mutations with a less severe initial impact would persist in the population for a longer period of time.

A permanent doubling of the current mutation rate would, however, after perhaps hundreds of generations, lead to a new equilibrium, with the incidence of genetic disease at about twice its current level. One of the biggest unknowns about predicting future effects of mutations is the impact of an accumulation of recessive mutations in the population. There is also little information to bring to bear on the question of the impact of an increase in mutations to the prevalence of common diseases.

Chapter 3

Kinds and Rates of Human Heritable Mutations

Contents

TABLES

Chapter 3

Kinds and Rates of Human Heritable Mutations

INTRODUCTION

The genetic code specified in DNA directs the transcription of RNA, which is translated into polypeptide chains; these chains, in turn, are assembled into proteins, which are the body's tools for the regulation of physiological, biochemical, and behavioral functioning. Mutations, transmissible alterations in DNA, can change the messages, causing alterations in RNA molecules and proteins. Some of these changes can lead to subtle abnormalities, diseases, and disabilities.

Environmental factors and specific interactions between environmental and genetic factors are thought to account for the majority of diseases or susceptibilities to disease. Mutations alone are thought to account for only a fraction of the current burden of disease. However, there is a group of disorders for which new mutations are the primary causes. Among these "genetic" diseases is a group of rare clinical conditions, "sentinel phenotypes," that occur sporadically in each generation; children with sentinel phenotypes are usually born to parents who do not have the same disease, indicating a new mutation that arose in a germ cell of one of the parents.

There are several other indicators of new mutations in human beings that do not rely on detecting diseases. The most widely used methods have been cytogenetic analysis of chromosomes and biochemical analysis of proteins.

The structure and number of chromosomes visible by light microscopy has been used to detect a certain class of heritable mutations. Gross chromosome abnormalities, often associated with high fetal and neonatal death rates, are identifiable with increasing precision by cytogenetic analysis. A large portion of these chromosome abnormalities arise anew each generation.

Studies of mutationally altered proteins have been done on a limited basis. The proteins most accessible for analysis are those circulating in blood; other body proteins are not routinely obtained for analysis. The method available for detecting abnormal proteins is electrophoresis. Heritable mutations have been detected by identifying electrophoretic variants of blood proteins. In addition, a combination of approaches has been used to estimate the mutation rate per nucleotide in genes coding for globin polypeptides, constituents of hemoglobin. These approaches to identifying mutations are discussed in this chapter.

The term "mutation rates" for particular types of mutations in specific regions of DNA is used frequently in this report. Mutation rates for different kinds of mutations, or corresponding to different overt effects of mutations, can be expressed as mutations per locus, per gene, per nucleotide, and per gamete. All of these indicate a specific type of mutation occurring *per generation*, reflecting mutations arising anew from one generation to the next. This chapter summarizes current knowledge about such mutation rates as they have been measured thus far; it is not yet known whether these rates apply also to other populations, to other gene regions, or to other types of mutations than the ones examined.

Rather than looking directly at germinal mutations—mutations in egg and sperm DNA—the methods described in this report focus on heritable mutations in liveborn offspring, a subset of all germinal mutations.[1] A larger subset of germinal mutations includes new mutations observed in offspring before birth. In recent years, cytogenetic techniques have been used to identify chromosome abnormalities in spontaneous abortuses. These investigations have shown that the frequency of new mutations in spontaneous abor-

[1]At present it is not feasible to examine egg or sperm DNA for genetic damage, although some information on major chromosome abnormalities can be obtained by cytogenetic examination of sperm.

tuses is much higher than in liveborn infants; the observed incidence of new mutations at birth only partially represents the "true" incidence of such heritable mutations in all conceptuses, and may give an incomplete picture of the potential risk from environmental exposures to mutagenic agents. Nevertheless, the most lasting medical and social consequences of new heritable mutations are derived from those mutations that are compatible with life at least through infancy and childhood.

SURVEILLANCE OF SENTINEL PHENOTYPES

The United Nations Scientific Committee on the Effects of Atomic Radiation (144) and the Committee on the Biological Effects of Ionizing Radiations (87) estimate that 1 percent of liveborn infants carry a gene for an autosomal dominant disease; in 20 percent of these cases (0.2 percent of livebirths) their disease is due to a new, or "sporadic," mutation that arose in the reproductive cells of one of their parents. Observing and recording the birth of infants with such sporadic conditions is the classical approach to identifying heritable mutations that lead to serious disease (36).[2]

A subset of the group of autosomal dominant and X-linked conditions, indicator conditions known as "sentinel phenotypes," may be particularly useful in identifying heritable mutations. Sentinel phenotypes are clinical disorders that occur sporadically, probably as the result of a single mutant gene. They are manifested at birth or within the first months of life, and usually require long-term, multidisciplinary medical care (83). In addition, sentinel phenotypes are associated with low fertility, so their appearance suggests mutations that have been passed not from *affected* parents to offspring, but instead, have been passed from *unaffected* parents to offspring as a result of *a newly arising mutation in a germ cell* of one of the parents of that offspring. Affected infants would have this mutation in the DNA of all their somatic cells and 50 percent of their germ cells.

Mulvihill and Czeizel (83) compiled a list, shown in table 1, of 41 genetic disorders (36 dominant and 5 X-linked) that satisfy the above criteria. Each of these phenotypes is individually rare, some of them occurring in only one in a million liveborn infants.

Since children with sentinel phenotypes have serious health problems associated with their conditions, they may first come to the attention of the family practitioner or pediatrician, and then be referred to subspecialists for a complete diagnosis. Clinical expertise in dysmorphology,[3] medical genetics, pediatric ophthalmology, and oncology may be needed to diagnose accurately the small numbers of children with sentinel phenotypes present among the vast majority of unaffected children and among children with phenotypically similar, but nongenetic conditions (called "phenocopies") (71,83,165).

Reliable incidence data are available only for 13 dominant and 5 X-linked sentinel phenotypes (166); only those that are manifested in infancy, rather than in childhood or adolescence,[4] are systematically recorded. Diagnostic records of sentinel phenotypes may be available in population-based registries of childhood cancers, birth defects, and genetic diseases, with initial entries based on birth certificates and later entries added as the conditions are diagnosed during infancy. After investigators rule out the possibility of a discrepancy between stated and biological parentage by genetic testing (166), sporadic cases of a sentinel phenotype are accepted as evidence for

[2]Rates of a wide variety of new heritable mutations leading to severe diseases have also been estimated by an "indirect method." This method is a theoretical approach, based on the loss of disease-causing genes from the population (due to reduced reproductive capabilities of individuals with these genes) and on the sporadic frequency of the genes in the population. The mutation rate necessary to maintain such genes in the population, despite their recurrent loss, is inferred from these observations. The types of populations and conditions suitable for this method are limited, and its precision is unknown (91,165).

[3]Physicians who specialize in diagnosing syndromes characterized by unusual physical form and stucture, for example, congenital malformations.

[4]Such disorders include achondroplasia, acrocephalosyndactyly, osteogenesis imperfecta, retinoblastoma, and Wilms' tumor (66).

Table 1.—Candidate Sentinel Phenotypes[1]

	Inheritance[a]
Phenotypes identifiable at birth:	
Achondroplasia	AD
Cataract, bilateral, isolated	AD
Ptosis, congenital, hereditary	AD
Osteogenesis Imperfecta type I	AD
Oral-facial-digital (Gorlin-Psaume) syndrome type I	XD
Incontinentia pigmenti, Bloch-Sulzberger syndrome	XD
Split hand and foot, bilateral atypical	AD
Aniridia, isolated	AD
Crouzon craniofacial dysostosis	AD
Holt-Oram (heart-hand) syndrome	AD
Van der Woude syndrome (cleft lip and/or palate with mucous cysts of lower lip)	AD
Contractural arachnodactyly	AD
Acrocephalosyndactyly type I, Alpert's syndrome	AD
Moebius syndrome, congenital facial diplegia	AD
Nail-patella syndrome	AD
Oculodentodigital dysplasia (ODD syndrome)	AD
Polysyndactyly, postaxial	AD
Treacher Collins syndrome, mandibulofacial dysostosis	AD
Cleidocranial dysplasia	AD
Thanatophoric dwarfism	AD
EEC (ectrodactyly, ectodermal dysplasia, cleft lip and palate) syndrome	AD
Whistling face (Freeman-Sheldon) syndrome	AD
Acrocephalosyndactyly type V, Pfeiffer syndrome	AD
Spondyloepiphyseal dysplasia congenita	AD
Phenotypes not identifiable at birth:	
Amelogenesis imperfecta	AD
Exostosis, multiple	AD
Marfan syndrome	AD
Myotonic dystrophy	AD
Neurofibromatosis	AD
Polycystic renal disease	AD
Polyposis coli and Gardner syndrome	AD
Retinoblastoma, hereditary	AD
Tuberous sclerosis	AD
von Hippel-Lindau syndrome	AD
Waardenburg syndrome	AD
Wiedemann-Beckwith (EMG) syndrome	AD
Wilms' tumor, hereditary	AD
Muscular dystrophy, Duchenne type	XR
Hemophilia A	XR
Hemophilia B	XR

[a]AD refers to autosomal dominant inheritance, XD refers to X-linked dominant inheritance, and XR refers to X-linked recessive inheritance.

SOURCE: J.J. Mulvihill and A. Czeizel, "Perspectives in Mutations Epidemiology 6: A 1983 View of Sentinel Phenotypes," *Mutat. Res.* 123:345-361, 1983.

a new germinal mutation originating in a germ cell of one of the parents (65).

In general, recording the incidence of sentinel phenotypes is not useful for monitoring a small population exposed to a suspected mutagen. Surveillance of sentinel phenotypes in large populations is potentially more useful for estimating the background mutation rates leading to dominant genetic disorders. Given the rarity of the individual sentinel phenotypes, the most reliable data for estimating mutation rates would be derived from the largest target populations with, for example, international cooperation among study centers (62,66).

Mutation Rates Estimated From Phenotypic Data

Data are available on the incidence of various sentinel phenotypes from different time periods in particular regions worldwide, though representation is incomplete. Crow and Denniston (22) summarized the most current data; these data have not been significantly updated since the 1940s and 1950s. The frequencies for any given phenotype are generally consistent between studies and the frequencies of occurrence of the different phenotypes vary over a thousandfold range, from 1 in 10,000 to 1 in 10 million depending on the particular disorder. The arithmetic mean of the rates for the different disorders is approximately 2 mutations per 100,000 genes per generation[5] (17,89,166). This average rate is often cited as the "classical" rate of heritable mutations.

[5]Mutation rates corresponding to sentinel phenotypes are expressed as the frequency of mutations per gene. It assumes one gene per disorder and one disorder per person and it is designated "per generation" because the mutations are detected in offspring of individuals whose germ cells have incurred mutation. In general, the mutation rate is defined as the incidence of a sporadic case of some indicator (e.g., a sentinel phenotype or a protein variant) divided by two times the total number of cases examined (e.g., newborn infants or protein determinations). The two in the denominator is to express the mutation rate, by convention, as the rate "per fertilized germ cell," instead of "per conceptus," which is made from two germ cells (162).

Limitations of the Phenotype Approach

Several sources of error may affect estimates of heritable mutation rates based on the incidence of sentinel phenotypes. The most important one results from the choice of phenotypes studied: phenotypes that occur in the population most often are the ones most likely to be recorded. Phenotypes that are so severe as to cause prenatal death, or so mild as to be indistinguishable from the normal variety of phenotypic traits, are difficult or impossible to include in these population surveys. It is not known how large a bias this selection occasions. In addition, the mutation rates corresponding to clinically recognizable phenotypes may not be representative of mutation rates for other traits.

Other problems contributing to an over- or underestimated mutation rate may include: 1) incomplete ascertainment among certain populations of liveborn infants; 2) mistaken identification of parentage, a significant issue since the number of cases of mispaternity may be high enough to exceed the number of true sporadic cases of sentinel phenotypes; and 3) the occurrence of phenocopies (phenotypically similar nongenetic causes of the same condition) and genocopies (recessive forms of the same phenotype inherited from unaffected carrier parents). In addition, particular phenotypes do not necessarily correspond to a unique, single gene (163); often, mutations in any one of several genes can lead to the same phenotype. This is not likely to influence the rate estimate significantly unless a large number of different mutant genes are involved, in which case the rate of mutation of the particular phenotype would be the sum of the rates of mutation at each of the alternative genes.

Surveillance for sentinel phenotypes is an important part of assessing the clinical impact of new mutations. The number of sentinel phenotypes available to study is limited, however, so that surveillance of these phenotypes illuminates only a fraction, albeit an important one, of this overall impact.

Individually, the sentinel phenotypes are quite rare. Keeping track of the frequencies of the different phenotypes over periods of time would require large-scale surveillance of huge numbers of infants over many years. A large team of medical specialists and a well-organized database would be needed to obtain complete and accurate ascertainment (65). Unfortunately, the practical difficulties in screening huge populations of newborns, diagnosing these rare disorders, and organizing and maintaining the data, make it difficult, though not impossible, to generate reliable data. The first such effort is currently being organized by the European Collaborative Study (62).

CYTOGENETIC ANALYSIS OF CHROMOSOME ABNORMALITIES

The adverse effects of chromosome abnormalities on human health are well established and provide sufficient medical reasons for screening infants for them. Data on the frequency of chromosome abnormalities is also useful in assessing the impact of detectable new chromosome mutations on the current generation.

Chromosome abnormalities are defined as either numerical (extra or missing whole chromosomes) or structural (deletions, insertions, translocations, inversions, etc., of sections of chromosomes). In liveborn infants, the presence of an extra or missing sex chromosome is often associated with physical, behavioral, and intellectual impairment. The presence of an extra autosome is even more detrimental: it is usually associated with severe mental and physical retardation and often with premature death (39,119).

There is no doubt that fetuses with recognized numerical chromosome aberrations are at a much higher risk for aborting spontaneously than fetuses without such defects. Such chromosome abnormalities contribute to a large proportion of spontaneous abortions and stillbirths, and in liveborn survivors may also contribute to repeated early abortions and fertility problems in adulthood (169). The significance to human health of structural rearrangements of the chromosomes is less well defined, but such defects have been associated with mental retardation, physical malformations, and a variety of malignant diseases (43). In addition, carriers of balanced translocations may produce offspring with unbalanced complements of DNA; these offspring may die in utero and may account for a high spontaneous abortion rate among such carrier parents.

It is believed that at least 5 percent of all recognized human conceptions have a chromosome abnormality, although this is a crude estimate because it is not corrected for factors which could influence the occurrence of such abnormalities (such as maternal age) or for factors which could affect the rate of embryonic or fetal death in some of the disorders (43). Numerical and structural chromosome abnormalities have been detected in 50 to 60 percent of recognized spontaneous abortions (data summarized in ref. 12), 5 to 6 percent of perinatal deaths (ibid.), and 0.6 percent of liveborn infants (42).

Chromosome abnormalities are useful for measuring heritable mutation rates since they are usually identified as new mutations, either because the defect causes virtual infertility or sterility (as in the case of numerical chromosome abnormalities) and therefore could not have been inherited from a parent with the same defect, or because the parents of an individual with a chromosome abnormality (e.g., a structural abnormality) have been shown not to carry the same mutation. Numerical chromosome abnormalities and some structural abnormalities are detectable with cytogenetic methods using current chromosome staining and banding techniques.

In general, little is known about the genetic mechanisms of either numerical or structural rearrangements. They appear to result from complex processes during gametogenesis in which various defects may lead to the same or different outcomes (18,119). Consequently, frequencies of the various cytogenetic abnormalities may correspond to a combined rate of several different mutational events or to a single mutational event.

Individuals with chromosome abnormalities can be identified on the basis of karyotypes (see ch. 2). However, since the vast majority of conceptuses with chromosome abnormalities such as trisomies and unbalanced structural rearrangements are spontaneously aborted, many before a pregnancy is recognized, the frequency of these mutations *at birth* is known to represent only a small fraction of the presumed frequency of these mutations *at conception*.

Data on the frequency of chromosome abnormalities in fetuses at the time of amniocentesis (generally done at 16 to 18 weeks gestation) provides information not available in newborn screening. In addition, recent progress in cytogenetic analysis of chorionic villi cells makes it possible to identify chromosome abnormalities in fetuses at 8 to 10 weeks gestation (116). This technique provides even earlier information on possible chromosome abnormalities, before fetuses with certain abnormalities (particularly trisomies and unbalanced structural rearrangements) would have aborted spontaneously. The analysis of fetal cells obtained at amniocentesis has provided valuable data on the frequency of chromosome abnormalities in the second trimester of pregnancy (12,45,159.) However, estimates of mutation rates from those measurements may be inflated unless certain biases are corrected. Women who elect amniocentesis—generally older mothers, couples who have had previous reproductive problems, and/or couples who may have been exposed to mutagenic agents (101) —are at increased risk for chromosome abnormalities in their offspring.

Cytogenetic surveys of consecutive liveborn infants have provided the most extensive data on chromosome mutation rates, although it is known that these data underestimate the "true" rate of new chromosome mutations. It should also be noted that the prevalence of sentinel phenotypes and biochemical variants in liveborn infants may also underestimate their "true" rate at conception, but little is known about selection against these mutations in the developing fetus (e.g., molecular repair of the defect, or spontaneous abortion of the fetus), in contrast to the documented evidence for selection against *chromosome* mutations in the fetus (135.) However, the frequency at birth is probably the best indicator of events with the highest cost in terms of the social and economic burden of these diseases.

Mutation Rates Derived From Chromosome Abnormalities

At least 10 major research efforts worldwide have attempted to determine the prevalence of chromosome abnormalities in liveborn infants. It is possible to estimate the rate of new mutations underlying these defects, but difficult to compare with rates of sentinel phenotypes, since chromosome mutations, by definition, are multigenic changes and sentinel phenotypes correspond to single gene mutations.

The combined data from studies surveying a total of 67,014 newborn infants for various periods between 1974 and 1980 in the United States, Canada, Scotland, Denmark, the U.S.S.R., and Japan (120) are shown in table 2. Two of the studies used newer banding techniques that identify more abnormalities than conventional staining methods used in the remaining eight studies.

For conditions that are associated with virtual infertility, complete sterility, or prereproductive death, it is often assumed that each case is the result of a new mutation (unless there is evidence for gonadal mosaicism[6]). Such conditions include the autosomal trisomies (e.g., Down syndrome) and the sex chromosome aneuploidies (e.g., Turner's syndrome).

For these defects, the apparent mutation rate observed in liveborn infants is calculated as the number of individuals identified with a given abnormality divided by the total number of individuals examined. This mutation rate is expressed as the number of mutations per generation. For conditions which could be the result of either new mutations or inherited mutations, such as the various balanced and unbalanced structural rearrangements, it is necessary to karyotype the parents of infants found with these chromosome abnormalities, and to exclude inherited mutations from calculations of the de novo mutation rate. For these conditions, the observed mutation rate is determined using the following formula:

$$\text{Mutation rate} = \frac{\text{Number of patients with abnormality} \times \text{Proportion who are likely to have de novo mutations}^{7}}{\text{Total number examined}}$$

The data show that the frequencies of chromosome abnormalities in liveborn infants range from 2 in 100,000 newborns per generation for inversions to 121 in 100,000 newborns per generation for Trisomy 21 (Down syndrome), the most common newly arising numerical chromosome abnormality found at birth. Overall, the numerical abnormalities are much more frequent than the structural abnormalities. However, the rates for balanced structural rearrangements may be low in these data, since only two of the surveys used more sensitive banding methods which are needed to identify these more subtle aberrations.

Limitations of Cytogenetic Analysis

Screening for chromosome mutations in newborn infants gives only a partial picture of their incidence. The prevalence of a particular genetic mutation at birth can be thought of as the result of the incidence of the mutation at conception interacting with the probability of survival of conceptuses with the mutation. Fetal survival, in turn, depends on individual fetal and maternal attributes, as well as on the interaction between the two (136). Some mutations in the fetus may affect fetal survival, whereas other mutations may have no effect on survival. Major chromosome aberrations, in particular, numerical abnormalities and most unbalanced structural rearrangements, are often incompatible with fetal survival and are associated with an increased risk of prenatal loss (45).

Balanced chromosome rearrangements, however, may not lead to excessive prenatal loss. For this reason, the balanced structural rearrangements may be a useful indicator of new germinal mutations observed in liveborn offspring. Since some of these infants identified as having a bal-

[6]Mosaicism refers to the presence of two genetically different cell populations in the same individual. This can result from a mutation during one of the earliest divisions in the zygote, a few days after conception, that led to the proliferation of a large number of cells that were derived from the original cell bearing the mutation.

[7]Such proportions are based on empirical data for each type of chromosome aberration. In this case, data from Jacobs, 1981 were used and are shown in table 2 in parentheses in the third column.

Table 2.—Prevalence of Chromosome Abnormalities at Birth

Chromosome abnormality	Total population	Number of new mutants[a]	New chromosome abnormalities per 100,000 newborns per generation
Numerical anomalies:			
Autosomal trisomies:[b]			
Trisomy 13	67,014	3	5
Trisomy 18	67,014	8	12
Trisomy 21	67,014	81	121
Male sex chromosome anomalies:			
47,XYY	43,048 males	43	100
47,XXY	43,048 males	42	98
Female sex chromosome anomalies:			
45,X	23,966 females	2	8
47,XXX	23,966 females	24	100
Balanced structural rearrangements:			
Robertsonian translocations:[c]			
D/D[d]	67,014	48(2/29)=3.3	5
D/G	67,014	14(2/11)=2.5	4
Reciprocal translocations and insertions	67,014	60(13/43)=18	27
Inversions	67,014	12(1/8)=1.5	2
Unbalanced structural rearrangements:			
Translocations, inversions, and deletions	67,014	37(7/16)=16	24

[a]See discussion in text for proportion of new mutants among all those identified with structural rearrangements.
[b]Trisomies refer to conditions in which there are 47 chromosomes (instead of the normal 46 chromosomes or 23 pairs of homologous chromosomes, indicated as 46,XX for females or 46,XY for males) because of the presence of an extra copy of one chromosome; three copies of this particular chromosome would be present instead of the normal two. *Autosomal* trisomies indicate an extra autosome, which includes any chromosome except one of the sex chromosomes.
[c]Rearrangements in which the long arms of two chromosomes fuse, resulting in one "hybrid" chromosome and the loss of the short arms of each chromosome.
[d]D and G refer to groups of chromosome numbers 13-15 and 21-22, respectively.

SOURCE: K. Sankaranarayanan, *Genetic Effects of Ionizing Radiation in Multicellular Eukaryotes and the Assessment of Genetic Radiation Hazards in Man* (Amsterdam: Elsevier Biomedical Press, 1982), pp. 161-164.

anced chromosomal mutation may have inherited the relatively subtle abnormality from one of their parents, it is necessary to determine whether the mutation observed in the child is actually a new germinal mutation or not.

Although it is not understood how bands observable under the microscope correspond to the structure and composition of DNA, mutations in DNA can cause a visible change in the banding pattern, particularly if such mutations involve a large section of a chromosome. Increased resolution and sensitivity of cytogenetic techniques, specifically the "high-resolution" banding methods (72), makes it possible to detect smaller and more subtle aberrations than was previously possible. Currently, routine cytogenetic methods can produce 200 to 300 bands (containing several hundred genes in each band) in a complete set of DNA, and the high-resolution methods can distinguish as many as 1,000 bands (containing about 100 or fewer genes per band) (59). With more smaller bands distinguishable, fewer genes are present per band, and a change in one gene has a greater likelihood of showing up as a change in the banding pattern.

At present, data available for large populations rely on only the routine staining and banding methods. Like the observations of sentinel phenotypes in newborn infants, cytogenetic methods focus on a subset of mutations, albeit those with serious clinical consequences. Even though chromosome abnormalities are not as rare as sentinel phenotypes, these two groups of abnormalities together probably account for only a small fraction of heritable mutations.

Mutations detectable by the methods discussed in the remainder of this chapter and in the following chapter do not necessarily have adverse clinical consequences. Their detection broadens the spectrum of identifiable mutations by including those that may not have an immediate adverse effect on the individual, and also offers more precise information about the nature of the genetic changes.

DETECTION OF BLOOD PROTEIN VARIANTS

In a single-gene, dominant genetic disease, if the causative mutant gene occupies one allele of the pair of genes and the normal form of the gene occupies the other allele, the one mutant gene is dominant over the normal gene and is sufficient to cause overt disease. *Dominant* diseases are said to be expressed in individuals who have a single dose, or are heterozygous, for the disease-causing gene. Autosomal *recessive* diseases, however, are clinically apparent only when the mutant gene is present at both alleles, or in the homozygous state; a single dose, the heterozygous state, can, for instance, destroy the activity of an enzyme or produce a defective gene product, but the corresponding normal gene supplies a sufficient amount of normal gene product to prevent overt clinical symptoms. It is difficult to infer the frequency of recessive genes by population studies of phenotypes since individuals with one copy of the recessive mutant gene are phenotypically indistinguishable from individuals with only normal genes for the particular protein. Biochemical means of identifying of heterozygotes, who are phenotypically normal, is one way of identifying new mutations that are expressed as recessive traits.

Biochemical studies on asymptomatic individuals with protein variants (indicative of single gene, recessive mutations) can be useful in expressing the cumulative, "invisible" effect of an increased mutation rate. Any protein in the body could be studied, but since peripheral blood is easily accessible, proteins from blood are most commonly used for these studies. Three major techniques for the detection and characterization of variant proteins have been used in studies of human populations:

1. electrophoresis of blood proteins,
2. quantitative and kinetic studies of enzyme deficiency variants from erythrocytes, and
3. molecular analysis of Hemoglobin M (HbM) and unstable hemoglobins.

One-Dimensional Electrophoresis

Electrophoresis separates proteins according to net molecular charge. A mixture of proteins is applied to a starch or acrylamide gel and exposed to an electric current. Each protein moves according to its own electric charge and separates from proteins with different charges. The subsequent addition to the gel of stains that bind to the proteins makes it possible to visualize the location of the proteins. Electrophoresis can be used to detect variant proteins in individuals and in their parents. The identification in an offspring of a variant protein that is absent from either parent is evidence for a new mutation. As in the other approaches, it is necessary to exclude the possibility of a discrepancy between stated and biological parentage so that only new, not inherited, mutations are considered.

Electrophoretic variants are detectable because a mutation results in a change in the net charge of a protein. On a theoretical basis it is expected that approximately one-third of amino acid substitutions produce a change in molecular charge of a protein, and there is also evidence that electrophoresis can detect changes in a molecule's configuration due to a nucleotide substitution. All in all, it is estimated that electrophoresis probably detects about 50 percent of nucleotide substitutions in expressed regions of coding genes (96).

The protein variants detected by electrophoresis are usually not associated with clinically recognizable problems. However, this technique has an important advantage for mutation research over the phenotypic approach; it measures changes at a level closer to the level of the mutational events and therefore may provide more accurate estimates of mutation rates. It also reflects more subtle genetic changes, broadening the spectrum of detectable genetic endpoints.

Mutation Rates Derived From Electrophoresis

Several large-scale studies to detect protein variants in human populations have been reported (excluding studies in Japanese atomic bomb survivors, which are described below). Neel, et al. (94), and Mohrenweiser (76) devised a large-scale pilot program to study mutation rates using placental cord blood obtained from newborn infants in Ann Arbor, Michigan (approximately 3,500 samples). They found no new heritable mutations in 36 different proteins in a total of 218,376 locus tests in Caucasian and 18,900 such tests in black infants.[8] Harris and colleagues (38) summarized data from studies in the United Kingdom in which 43 different loci coding for blood proteins were analyzed, and no new mutations were found in 133,478 locus tests.[9] Altland and colleagues (5) examined blood from filter paper submitted originally for screening for phenylketonuria in approximately 25,000 newborn infants in West Germany. They identified one putative new mutation among five different hemoglobin proteins in approximately 225,000 locus tests.

Feasibility of Electrophoresis

The availability of human blood for analysis of variant proteins makes electrophoresis feasible for studies of large populations. In addition, it permits comparisons of the frequency of occurrence of variant proteins between humans and other species, since comparable proteins are usually associated with homologous gene sequences (93). Perhaps the most important advantage of this approach is that it permits the study of a defined set of genes expressing functional proteins and offers more precise information on the number of genes involved in the biochemical observations made and on the effect of mutation on some proteins (90).

However, electrophoretic analysis can provide information only on mutations in genes that are transcribed into RNA and subsequently translated into proteins, and only for mutations that alter the protein's net molecular charge or configuration. It cannot provide information on mutations in nontranscribed regions of the DNA, which comprise the larger portion of the genome (165).

Since the baseline frequency of new protein variants is still poorly defined, it is difficult to estimate the size and scope of a population study necessary to detect an increase over a period of time or to detect differences in rates between two populations. One-dimensional electrophoresis is limited by the number of proteins that can be examined for variants, approximately the same number of different phenotypes included in the list of sentinel phenotypes. However, one-dimensional electrophoresis in a large population is a more straightforward approach to studying new heritable mutations than is population surveillance for sentinel phenotypes.

Vogel and Altland (164) estimated that analysis of two populations of 10 million individuals each would be needed to have a 95-percent chance of detecting a 10-percent increase in spontaneous mutation rates. This calculation is based on a theoretical rate of 1 mutation per 1 million genes per generation and 50 loci tested per individual in the study. The population size could be reduced if the proteins that were examined mutated more frequently or if detection of only large increases in the mutation rate (e.g., 50 percent) were acceptable, or if many more proteins could be examined from each individual tested.

Quantitative and Kinetic Studies of Enzyme Deficiency Variants

Mutations that result in variant proteins with greatly reduced or no biological activity, when present in the heterozygous state, are not readily detectable in electrophoretic studies. Such "enzyme deficiency variants," or "nulls,"[10] can be identified through automated biochemical tests that measure enzyme activity in red blood cells. These studies increase the detectable range of mutational events.

Loss of function of a protein may result from:

- the absence of a gene product,
- the presence of a protein which is nonfunctional catalytically, or
- abnormally unstable enzymes (75).

The types of mutations causing these abnormalities include chromosome rearrangement or loss, deletions, frameshift mutations, and nucleotide

[8]The number of locus tests takes into account the number of proteins examined, the number of gene loci represented by each such protein, and the number of individual samples obtained.

[9]The number of individuals tested varies depending on the enzyme examined, but for some of the enzyme tests, more than 10,000 individuals were tested.

[10]These mutations are characterized by the presence of an enzymatically inactive protein, or by an absence of a protein. These can be caused by gene deletions, nucleotide substitutions in critical locations in the gene, mutations in the "start" and "stop" codons for the protein, or even by nucleotide substitutions at splicing junctions for the introns.

substitutions in coding or noncoding regions of DNA. The range of genetic events underlying the production of enzyme deficiency variants is thought to be larger than for electrophoretic mobility variants, since mutations in noncoding regions (e.g., intervening sequences, flanking regions, etc.) as well as in coding regions could be detected (74).

Four studies have provided data on the frequency of enzyme deficiency variants. The frequencies of such variants have been measured in: 1) placental cord blood samples from approximately 2,500 individuals in Ann Arbor, Michigan (74,75); 2) blood samples obtained from participants of the studies in Japan coordinated by the Radiation Effects Research Foundation (122); 3) blood samples obtained from 3,000 hospitalized individuals in West Germany (30); and 4) blood samples obtained from Amerindians of South and Central America (78). No new mutations were found in any of these studies. However, rare but not new enzyme deficiency mutations that have been inherited over many generations have been measured at an overall frequency of 2 variants per 1,000 determinations in a total of approximately 110,000 locus tests. If new mutations producing enzyme deficiency variants occur, as predicted, no more than twice as often as new mutations leading to electrophoretic variants, the lack of new mutations in these 110,000 locus tests is not inconsistent with electrophoretic data.

Molecular Analysis of Hemoglobin M and Unstable Hemoglobins

Stamatoyannopoulos and Nute (131) calculated mutation rates for certain nucleotides in the globin genes, which, when mutated, produce autosomal dominant disorders expressed as chronic hemolytic anemia or methemoglobinemic cyanosis. Their data provided the first direct measurements of heritable mutation rates for particular *nucleotides* in human DNA. Although these data are limited to the specific characteristics of the globin genes, they suggest the kind of information which may be possible to obtain for other genes as the genetic basis for various single gene disorders is discovered. Even if the mutation rates derived from these studies of the globin genes are not representative of rates of mutation at other loci, the hemoglobin system described below provides a model for combining epidemiologic, clinical, and molecular information to estimate heritable mutation rates in human populations.

Structure and Function of Hemoglobin

Hemoglobin molecules, which exist in a number of slightly different forms, are the universal carriers of oxygen in the blood from the lungs to all cells of the human body. Different mixtures of hemoglobins are present at all stages of development from embryonic to adult, although a particular type of hemoglobin tends to predominate at any one stage. In addition, hundreds of abnormal hemoglobins have been identified, ranging in clinical severity from benign through serious hematologic disorders. Human hemoglobin is a protein consisting of four separate globin polypeptide chains and four iron-containing heme groups. In adults, the predominant form of hemoglobin is hemoglobin A (HbA), composed of two alpha-globin chains and two beta-globin chains each joined to a heme group.

Hemoglobin's primary function of binding, carrying, and releasing oxygen is closely regulated by several factors, including the maintenance of a precise three-dimensional conformation of the hemoglobin molecule, the composition and balance of globin chains, and the distribution of ionic charges in and around the molecule. Mutations which result in amino acid substitutions at particular sites in the globin polypeptide chain can markedly alter these specific functional properties of the hemoglobin molecule.

Two types of abnormal hemoglobin, HbM and unstable hemoglobin, usually occur sporadically as a result of a new mutation and are expressed as dominant mutations.[11] Stamatoyannopoulos and Nute (131) described the genetic changes leading to these particular abnormal hemoglobins and their incidence in human populations. From this

[11]A heterozygous individual, having one mutant globin gene and one normal allele, would have clinically recognizable symptoms of the disease. This is in contrast to other, more common hemoglobin diseases, such as sickle cell anemia and thalassemia, where the heterozygote is asymptomatic and the disease occurs only when an individual is homozygous for the variant gene.

information, they were able to calculate the rate of mutation leading to these abnormal hemoglobins.

Unstable Hemoglobins

Various mutations in the globin genes can lead to an altered sequence of amino acids in the globin chains and to a major change in the three-dimensional structure of the hemoglobin molecule. Often these changes create a less stable hemoglobin molecule than its normal counterpart. "Unstable hemoglobin" can result in the hemoglobin precipitating within the red cell, leading to membrane damage and premature destruction of the red cells by the liver and spleen (165).

The degree of clinical severity of the unstable hemoglobin depends in part on the site at which the amino acid replacement occurs and on the properties of the amino acid that is substituted (37). However, manifestations of the disorder vary from mild hemoglobin instability which may not be clinically significant, to severe hemoglobin instability, which causes chronic hemolytic anemia.

Unstable hemoglobin is diagnosed by isolating and characterizing the abnormal hemoglobin, and by determining its molecular stability. Since the amino acid sequence of normal hemoglobin is known, and the sequences for some abnormal hemoglobins have been determined, the genetic mutation underlying these unstable globins have been identified.

Hemoglobin M—The Methemoglobinemias

A blue, cyanotic appearance is the chief clinical manifestation of methemoglobinemia, caused by HbM. Cyanosis occurs whenever a significant proportion of the hemoglobin in the circulating red blood cells is not carrying oxygen, and in this case it is because of an inability of the hemoglobin to bind and carry oxygen (37).

Five different mutations in the globin gene which lead to the production of HbM have been identified. All of these are expressed as dominant phenotypes. The common characteristic of these mutations is that each results in substitution of an amino acid in the globin peptide where the heme groups are attached. (These mutations are specific substitutions of histidine for tyrosine on the alpha-globin chain [at positions 58 and 87] or on the beta-globin chain [at positions 63 and 92], or by substitution of valine for glutamic acid on the beta-globin chain [at position 67].) By altering the site of attachment of heme on the globin chains, these mutations change the three-dimensional organization of the entire hemoglobin molecule and interfere with its ability to combine with oxygen (100).

Measurement of Mutation Rates in the Globin Genes

Using published data on the number of cases of hemoglobin disorders from 10 countries (134), Stamatoyannopoulos and Nute (131) determined the number of children with HbM and unstable hemoglobin among an estimated number of livebirths in the defined populations. They recorded a total of 55 cases of de novo hemoglobin mutants, each of which derived from substitutions of single nucleotides in the coding portions of the globin genes: 40 children with unstable hemoglobin (all beta chain mutations) and 15 with HbM (10 beta chain and 5 alpha chain).[12] The requisite paternity data, also from published information on these mutations, had been collected on 19 of the 55 cases which were used to calculate mutation rates (131). Their data are shown in table 3.

The mutation rates derived from these data are expressed in mutational events per alpha- or beta-gene nucleotide per generation, and range from 5.9×10^{-9} to 19×10^{-9} for alpha- and beta-gene nucleotides, respectively. The rate for all beta-gene variants together is 7.4×10^{-9} per nucleotide per generation. These estimates correspond to mutation rates per globin chain *gene* per generation of 2.6×10^{-6} to 8.3×10^{-6}. The rate per gene is obtained by multiplying the rate per nu-

[12]The number of alpha chain mutations detected was less than expected, given the existence of two alpha-globin genes per beta-globin gene in an individual. A possible reason for this may be that alpha-chain defects would produce physiological abnormalities at an earlier stage of fetal development than would beta-chain abnormalities, since the alpha gene is activated much earlier, and may be causing a greater loss of fetuses early in gestation.

Table 3.—Rates of Mutation Leading to HbM and Unstable Hb Disorders

Type of disorder	Mutation rate/ nucleotide/ generation[a]	Mutation rate/ globin gene[b]
Unstable Hbs . . .	5.9×10^{-9}	2.6×10^{-6}/438 nucleotides
Alpha^M variants .	$10.0\times^{-9}$	4.2×10^{-6}/423 nucleotides
Beta^M variants . .	18.9×10^{-9}	8.3×10^{-6}/438 nucleotides
All beta-chain variants	7.4×10^{-9}	3.2×10^{-6}/438 nucleotides

[a]The mutation rate per nucleotide is based on the rate of appearance of the particular disorder (number of mutants observed divided by 2 multiplied by the number of births in the generation(s) included). This is the average mutation rate for the particular disorder expressed per globin gene affected per generation. To calculate the mutation rate per nucleotide per generation, Stamatoyannopoulos and Nute took the frequency of each mutant divided by 2 (for two beta genes in a genome) multiplied by 3 (since a given nucleotide can be substituted by any one of three other nucleotides), then divided that product by the number of different substituted nucleotides actually observed in their data.

[b]The estimated mutation rate per alpha-globin and beta-globin gene was calculated as the mutation rate per nucleotide multiplied by the number of nucleotides constituting the coding portion of the corresponding gene (423 in the alpha-globin gene and 438 in the beta-globin gene).

SOURCE: G. Stamatoyannopoulos and P.E. Nute, "De Novo Mutations Producing Unstable Hbs or HbsM. II: Direct Estimates of Minimum Nucleotide Mutation Rates in Man," *Hum. Genet.* 60:181-88, 1982.

cleotide by the number of nucleotides in the coding portion of the gene.

The rates calculated from these data are likely to be minimum rates corresponding to these disorders. The diagnosis of these hemoglobin disorders requires structural and functional analyses of the aberrant proteins, and not all cases would be identified initially by their phenotypic abnormalities; some degree of underascertainment is inevitable in these data. This is particularly true in the data for the unstable hemoglobins, where clinical manifestations of the genotype range from mild to severe. The more severe cases are likely to be identified and reported in the literature.

Overall, the order of magnitude of the estimates for gene mutation rates, 1 mutation in 1 million genes per generation, is consistent with that of the estimates of the spontaneous heritable mutation rate made by Neel and colleagues in the Japanese populations that comprised the control group in their ongoing study of the genetic effects of the atomic bombs (see below). This suggests that even though the rates derived from these data are based on mutations observed in only alpha- and beta-globin genes, they are not very different from mutation rates in other expressed genes.

STUDIES OF POPULATIONS EXPOSED TO ATOMIC RADIATION

The first demonstration that ionizing radiation could induce genetic mutation is usually credited to Müller (81) for his work with Drosophila, and since then, many scientists have examined the nature and frequency of induced mutations in various species of animals, including whole mammals (120). There is an extensive body of experimental data that suggests that exposure to radiation and to certain chemicals can induce mutations in mammalian germ cells. In humans, exposure to ionizing radiation is known to cause *somatic* mutation, and is suspected to enhance the probability of germinal mutation (165). *To date, however, the available methods have provided no documented evidence for the induction (by chemicals or by radiation) of mutations in human germ cells.*

The single largest population at risk for induced heritable mutations is the group of survivors of the atomic bombs dropped on Hiroshima and Nagasaki in 1945. Survivors of these bombs are estimated to have received doses of radiation considered to be "biologically effective"; in experiments with mutation induction by acute exposure to radiation in mammals, similar kinds and doses of radiation were sufficient to cause phenotypically apparent mutations in offspring (and presumably other nonexpressed mutations as well). On the basis of this kind of experimental data, the assumption was made that germinal mutations could have been induced in people exposed to the radiation.

Efforts to detect genetic consequences of the atomic bombs detonated in Hiroshima and Nagasaki have been in progress continuously since 1946, sponsored by the Atomic Bomb Casualty Commission and by its successor in 1977, the Radiation Effects Research Foundation. These studies have attempted to determine whether germinal mutations are expressed as heritable mutations in the survivors' offspring, particularly as "untoward pregnancy outcomes," as single gene disorders, as chromosome abnormalities, or most recently, as abnormal blood proteins.

Early Epidemiologic Studies

The earliest clinical studies, as well as the continuing mortality surveillance study, sought to identify a variety of health problems in the offspring of exposed atomic bomb survivors. These studies consisted of epidemiologic surveys of 70,082 newborn infants born in Hiroshima and Nagasaki between 1948 and 1953, including about 38,000 infants for which at least one parent had been proximally exposed to the radiation (defined as being within 2,000 meters of the hypocenter, while distally exposed was defined as being greater than 2,500 meters from the hypocenter or not exposed at all). Stillbirths, infant mortality, birthweight, congenital abnormalities, childhood mortality, and sex ratio were examined. An "untoward pregnancy outcome" was defined as a stillbirth, a major congenital defect, death during the first postnatal week of life, or a combination of these events. The investigators found no statistically significant, radiation-related increase for any "untoward pregnancy outcomes" (123). One reason for this finding may be that these health problems each have some genetic basis, but that all are influenced by a variety of nongenetic factors as well, some of which may have obscured an effect of genetic mutation. In addition, a continuing study (the F1 Mortality study) of mortality among children born between 1946 and 1980 has not demonstrated a relationship between radiation exposure of the parents and mortality in the offspring (56).

Cytogenetic Studies

Since 1967, Awa and colleagues have screened the children of survivors of the atomic bombs for cytogenetic abnormalities to determine whether exposure of the parents' gametes to the radiation caused a higher frequency of chromosome abnormalities in the offspring. After first reporting a higher frequency of sex-chromosome abnormalities among 2,885 children of exposed parents compared with 1,090 children of the control group of parents, Awa (6) concluded that a mutagenic effect of the radiation was not demonstrated in these data. The total frequency of chromosomally abnormal children of exposed parents (0.62 percent) was higher than in the control group (0.28 percent), although the difference was not statistically significant, and was similar to the frequency of such abnormalities observed in several studies of unselected, consecutive newborn populations (120).

Results of their more recent analysis of the data showed that 0.52 percent of the offspring of exposed parents were determined to have an abnormal chromosome constitution, compared to 0.49 percent of the offspring of the control group of parents, which suggests that there is no significant difference between the two groups of offspring in frequency of autosomal balanced and unbalanced rearrangements and of sex-chromosome numerical abnormalities (8). This study did not provide information on chromosome abnormalities which are associated with high mortality rates, such as most of the unbalanced structural rearrangements and numerical chromosome abnormalities involving the autosomes.

Biochemical Analyses

Beginning in 1976, one-dimensional electrophoresis was used to look for mutations affecting proteins found in red blood cells and blood plasma. Rare variants of these proteins, differing by a single constituent amino acid, are taken to represent mutations which occurred in the germ cells of one of the parents if the variant is detected in a child and is absent from both parents. Classification as a "probable mutation" is made after inquiry into possible errors of diagnosis and of stated parentage (124).

Neel and colleagues (93,96,121) analyzed 28 enzyme and serum proteins from blood samples obtained from children who were identified in the F1 Mortality study and who were born between 1959 and 1975 to survivors of the atomic bombs. Over 10,000 children were examined in each group of exposed and unexposed parents. These children were also participants in the cytogenetic studies, and the biochemical analyses made use of the same blood samples that were used for the chromosome analyses. To date, two probable mutations have been identified among offspring of proximally exposed parents (estimated to have received an average conjoint gonadal exposure of approximately 600mSv [60 rem[13]]) in 419,666 locus tests, while 3 new mutations have been found among 539,170 locus tests of unexposed individuals. Although the number of new mutants is small, a crude mutation rate can be calculated from these data. The rate of mutation in the exposed group is approximately 5 mutations in 1 million genes, and the rate in the unexposed group is 6 mutations in 1 million genes per generation. Such rates have wide margins of error and are not statistically different (96).

In 1979, the tests of electrophoretic variants were expanded to include assays for reduced activity variants of certain red cell enzymes.[14] The presence of such variants suggests a deletion in the DNA, producing "null" mutations which would not be identified by one-dimensional electrophoresis. No such variants were found among 11,852 locus tests of 10 erythrocyte enzymes in children of exposed and unexposed parents (122).

[13]The doses of radiation that the survivors are estimated to have received are currently being re-evaluated in light of new information concerning the quantity and quality of radiation released by the two bombs (125). The revised estimate of exposure will probably be lower than the current estimate of 60 rem.

[14]An enzyme level three standard deviations below the mean and/or less than or equal to 66 percent of normal was defined as a deficiency variant.

Mutation Studies in Populations Exposed to Radiation in the Marshall Islands

Neel and his colleagues (93) reported results of electrophoretic studies of blood proteins from offspring of parents exposed to radioactive fallout from the Bravo thermonuclear test explosion at Bikini in the Marshall Islands in 1954. No variant proteins indicating new mutations were found in 1,897 locus tests of 25 proteins measured in children of unexposed parents, and in 1,835 such tests in children of exposed parents.

CONCLUSIONS

This chapter has described several currently feasible approaches for determining rates of heritable mutations. Each of the methods has its particular problems. The genetic defects underlying the sentinel phenotypes are largely unknown. These phenotypes may result from mutations within genes at one or several genes, and may be complicated by the presence of phenotypically similar but nongenetic traits, all serving to bias estimates of their frequency upwards and to create a wide margin of error. Even with a collection of sentinel phenotypes that can be reliably identified in newborn infants, population surveys of a sufficient size are difficult to perform. To detect a small or moderate *increase* of sentinel phenotypes, consecutive newborns of entire national populations would have to be monitored thoroughly for many years, perhaps as part of an ongoing surveillance program for birth defects and childhood cancers.

The routine staining and banding of chromosomes is the most straightforward of the current methods, but they allow only major chromosomal changes to be identified. More advanced banding techniques, which are not performed routinely, allow smaller chromosome abnormalities to be detected in some cases.

Among the current methods described here, the detection of electrophoretic protein variants comes closest to mirroring precise changes in DNA. However, one-dimensional electrophoresis is limited by the relatively small number of proteins it can examine and by the fact that it does not readily detect the absence of gene products, caused by deletions or by specific nucleotide substitutions. Detection of enzyme deficiency variants, although still in early stages of development, could complement electrophoresis by detecting loss-of-activity variants. In addition, large population surveys are required to generate statistically significant results, and many more paternity tests are likely to be needed as a result of the biochemical approach compared to the phenotypic and cytogenetic methods. (Since the protein variants are not associated with reduced fertility, as are most of the sentinel phenotypes and chromosome abnormalities, only a fraction of the protein variants will be new mutations.)

An interesting example of an approach for detecting mutations in a single gene or in a group of genes is provided by the analysis of mutations underlying unstable hemoglobins and HbM. Analyses of other accessible and well-studied genes may help to generate better estimates of mutation rates per nucleotide and per gene on the human genome, although it is not known how representative any one gene is of the remainder of the DNA. Moreover, characterizing the mutations at the molecular level provides important information, especially where specific mutagens are suspected to have caused the mutations.

The different approaches have led to various estimates of heritable mutation rates in human beings (see table 4). Sentinel phenotypes, representing the occurrence of a single dominant mutant gene, are detected in about 2 individuals, on average, per 100,000 examined per generation. Individual types of chromosome abnormalities are up to 50 times more frequent than the average frequency of sentinel phenotypes. Analyses of the mutant hemoglobins studied by Stamatoyannopoulos and Nute (131) give estimated spontaneous mutation rates of about three mutations per 1 million genes per generation. The rate measured in one-dimensional electrophoresis of blood proteins is also about three mutations per 1 million genes per generation, a figure based on the examination of over 1 million locus tests involving 30 to 40 different proteins.

The different approaches currently available for detecting heritable mutations (listed in table 5) in human beings generate data that are generally not comparable. Each set of data reflects the nature of the approach itself, not necessarily the nature of the different genes examined or their spontaneous mutation rates. It is particularly difficult to compare directly the results of studies of mutation rates based on clinical syndromes with biochemical studies. Measurements from each of the current methods can be used to generate a separate mutation rate and, for the time being, those rates can be viewed as baselines for the frequency of various kinds of human mutations and as reference points for some of the most severe effects of heritable mutations.

Table 4.—Results of Studies To Detect Heritable Mutations by One-Dimensional Electrophoresis in Human Populations Not Known To Be at High Risk for Germinal Mutations

Population	Locus tests	Mutations
United States[a]: Caucasian	218,376	0
Black	18,900	0
United Kingdom[b]	133,478	0
West Germany[c]	225,000	1
Central and South America[d]	118,475	0
Japan[d]	539,170	3
Marshall Islands[e]	1,897	0
Total	1,255,296	4

Overall "spontaneous" mutation rate = 4 mutations/1,255,296 locus tests is equivalent to approximately 3 mutations/1 million genes per generation.

SOURCES:

[a]J.V. Neel, H.W. Mohrenweiser, and M.H. Meisler, "Rate of Spontaneous Mutation at Human Loci Encoding Protein Structure," *Proc. Natl. Acad. Sci. USA* 77(10):6037-6041, 1980; H.W. Mohrenweiser, 1986, unpublished data, cited in Neel, et al., 1986.

[b]H. Harris, D.A. Hopkinson, and E.B. Robson, "The Incidence of Rare Alleles Determining Electrophoretic Variants: Data on 43 Enzyme Loci in Man," *Ann. Hum. Genet. Lond.* 37:237-253, 1974.

[c]K. Altland, M. Kaempfer, M. Forssbohm, and W. Werner, "Monitoring for Changing Mutation Rates Using Blood Samples Submitted for PKU Screening," *Human Genetics, Part A: The Unfolding Genome* (New York: Alan R. Liss, 1982), pp. 277-287.

[d]J.V. Neel, C. Satoh, K. Goriki, et al., "The Rate With Which Spontaneous Mutation Alters the Electrophoretic Mobility of Polypeptides," *Proc. Nat. Acad. Sci. USA* 83:389-393, 1986; J.V. Neel, unpublished data.

[e]J.V. Neel, H. Mohrenweiser, S. Hanash, et al., "Biochemical Approaches To Monitoring Human Populations for Germinal Mutation Rates: I. Electrophoresis," *Utilization of Mammalian Specific Locus Studies in Hazard Evaluation and Estimation of Genetic Risk*, F.J. deSerres and W. Sheridan (eds.) (New York: Plenum Press, 1983), pp. 71-93.

Table 5.—Spontaneous Mutation Rates of Genetic Abnormalities in Human Populations

Genetic anomaly	Mutations/ generation	References
Sentinel phenotypes	2 mutations per 100,000 individuals	(166)
Chromosome abnormalities	2 to 121 mutations per 100,000 individuals	Table 3-2
Electrophoretic protein variants	3 mutations per million genes	Table 3-4; (96)
Analysis of unstable hemoglobins	3 mutations per million globin genes	Table 3-3; (131)

SOURCE: Office of Technology Assessment.

The different approaches currently available reflect the various forms that mutations and their consequences take. *Each of the current approaches to studying mutations is generally limited to particular kinds of mutations or to a particular set of outcomes.* The cytogenetic and biochemical approaches discussed are limited to detecting a small subset of mutations that occur in human DNA. Identification of sentinel phenotypes and chromosome abnormalities limits the observations to a particular set of outcomes, mostly in the severe end of the spectrum. The available data are based on mutations observed in a small fraction of the genes, about 40 to 50, out of a total of about 50,000 to 100,000 in the human genome.

Aspects of the current methods that limit their scope also make them inefficient. The proportion of all mutations that these methods can identify is relatively small and the limited number of suitable genetic endpoints (acceptable sentinel phenotypes, chromosome abnormalities, and protein variants) makes it difficult to sample more than a few mutational events per individual. As a result, huge populations of "exposed" and "unexposed" people would be needed to find out whether the exposure resulted in excess heritable mutations (71). Not surprisingly, perhaps, there is no documented evidence to date for a meaningful difference in rates for any type of mutation between two populations (or in the same population at different times).

Sentinel phenotypes and cytogenetic analyses provide a highly selective approach to some of the worst possible consequences of new mutations. However, they do not permit more than a partial view of the broader consequences of mutations in human beings. Detailed molecular, physiological, and clinical information on the kinds and rates of new mutations will eventually be needed to assess the immediate or potential impact of new mutations.

Chapter 4

New Technologies for Detecting Heritable Mutations

Contents

TABLE

FIGURES

Chapter 4

New Technologies for Detecting Heritable Mutations

INTRODUCTION

New developments in molecular biology have opened the way for direct examination of large portions of human DNA to detect kinds and rates of mutations. Until recently, it has been possible only to infer the occurrence of genetic damage from indirect, limited, and often imprecise observations (see ch. 3). In principle, these new technologies would allow samples of DNA from people exposed to potential mutagens to be examined for precise genetic damage.

Little is known about the nature and frequency in humans of the majority of mutational events, since previous information was available only on DNA sequences that could be selected on the basis of gene expression. Until recently, human mutation was studied in fine detail only in particular genes (e.g., *hprt* and globin genes) that could be fished out of genomic DNA by using some gene-specific probe, or in genes whose products (mRNA, proteins, etc.) could be isolated and examined.

In this chapter, new methods are described for examining proteins and for examining large regions of genomic DNA (see table 6), regardless of their function or location in the genome and without devising selection methods for specific known genes. These new technologies are currently under development or being considered by the scientific community. Most of them were explored as potentially feasible technologies at a workshop cosponsored by the International Commission for Protection Against Environmental Mutagens and Carcinogens (ICPEMC) and the U.S. Department of Energy, in December 1984.

Table 6.—Current and New Methods for Detecting Human Heritable Mutations

Indicators of mutation	Method of detection
Current methods:	
Sporadic genetic disease	Sentinel phenotypes
Numerical or structural chromosome abnormalities	Cytogenetic analysis
Variant blood proteins	1D-electrophoresis Quantitative enzyme analysis 2D-electrophoresis
New methods:	
Altered DNA sequences	DNA sequencing Restriction fragment length polymorphisms 1D-denaturing gel electrophoresis 2D-denaturing gel electrophoresis Ribonuclease cleavage Subtractive hybridization Pulsed field gel electrophoresis

SOURCE: Office of Technology Assessment.

Several different approaches are used in these new technologies. Two-dimensional electrophoresis of proteins follows from the techniques of one-dimensional electrophoresis and incorporates new methods of analyzing protein spots on the gel. In several of the DNA methods, restriction enzymes are used to fragment genomic DNA before sizing and sequence analyses are performed. Other technologies hinge on the production of hybrid DNA molecules, or heteroduplexes, and on gradient denaturing gels that are used to identify heteroduplexes with base pair changes. Another method devises a way of separating larger fragments of DNA than were previously possible to separate. These new methods are discussed below.

SHORT DESCRIPTIONS OF THE NEW TECHNOLOGIES

Two-Dimensional Polyacrylamide Gel Electrophoresis

A development following from the experience with one-dimensional electrophoresis for protein variants (see ch. 3) is two-dimensional polyacrylamide gel electrophoresis, "2-D PAGE," the only technique described in this chapter that is currently in use in studies of human beings. With this technique, proteins are separated first by isoelectric focusing on the basis of their molecular charge, and then by electrophoresis on the basis of their molecular weight. From 500 to 1,000 proteins may be visible from a single sample on one gel, and about 100 of these may be sufficiently clear to detect rare variants in a child that are not present in the parents (92).

To search for new mutations in proteins using 2-D PAGE, blood samples are taken from child/mother/father trios. The program that has had most experience with this technique for mutation research, the University of Michigan Medical School, collects cord blood from newborns and parental blood samples at the time of the child's birth. Each blood sample is fractionated into six parts: platelets, plasma, nonpolymorphonuclear cells, polymorphonuclear cells, erythrocyte (red blood cell) membranes, and erythrocyte cytosol (the contents of the cell) (95). Each fraction is treated and analyzed separately.

Separation in the first dimension is by the charge of the protein molecules, qualitatively similar to the separation using one-dimensional electrophoresis. Each sample is run on a long, thin (1 to 2 mm in diameter) cylinder of polyacrylamide in a hollow glass tube using a technique called isoelectric focusing. The separation time can be speeded up or slowed down depending on the voltage applied. After the first separation, the gel, which looks like a clear noodle, is "extruded" from the tube.

In the second separation, the "noodle" is laid across the top of a gel about 7 inches square, that is held between two glass or plastic plates. A current is passed through the gel, and the protein molecules migrate down through the gel at a speed that is proportional to their size. A final step renders the proteins visible by autoradiography (for proteins that can be isotopically labeled) or by stains such as Coomassie Brilliant Blue and silver- or nickel-based stains.

The trained eye or a computer program detects patterns of spots on the gel, though most of the stained areas are too crowded or indistinct for drawing conclusions about specific proteins. Only spots in relatively clear areas and of a sufficient intensity are chosen for scoring. The analysis entails comparing the parent/child trios for differences. No two gels are run in precisely the same way, so it is often difficult to superimpose a set of three gels over one another and to look for differences. While protein spots bear the same relationships to each other on different gels, they routinely appear at slightly different coordinates.

For each type of sample mentioned above, a standard constellation of protein spots is scored for genetic variation. A mutant protein can appear on the gel as a spot for which there is no corresponding spot on parental gels. The gels can be interpreted visually, which is the traditional method, and currently the most accurate, but computer-assisted interpretation is desirable. Two steps in analysis lend themselves to computerization. First, and now routinely done, the protein spots can be "read" by computer for location and intensity. The second step is computerized comparison of different gels. Two approaches for comparison are being developed in different laboratories, each of which will probably be useful on a "production scale" in the near future (92). Eventually, computers will be relied on to screen gels, and to eliminate from further consideration the overwhelming majority of parent/child trios with no variants.

In comparison to one-dimensional protein separation, a relatively small number of parent/child trios have been analyzed with 2-D PAGE, and no heritable mutations have been found. However, the results of these analyses have revealed a wide array of largely unknown proteins, and distinct differences in the spectrum of proteins from different cell types. The range of electrophoretic variants of a number of these proteins have been characterized in plasma (104) and in erythrocyte lysate (105). In plasma, the variation is substantially higher than has been reported in other blood fractions.

The advantage of 2-D PAGE is that it can be done now without a major developmental breakthrough and without a great initial cost, though the running expenses are significant. It may be some years before any of the DNA analytic methods catch up to protein analysis in practical application. The future of protein analysis for detecting mutations after DNA methods become routine is unclear, however.

There are technical limitations of the 2-D PAGE technique. First, the technique itself is technically demanding. Second, the number of proteins that can be scored on each gel is limited by the degree of separation possible on plates that are of a practical size.

As with most of the other techniques, only a certain spectrum of mutations is detectable with 2-D PAGE. Chromosome rearrangements in general are not revealed through electrophoretic variants. This technique has its greatest strength in detecting point mutations and larger deletions in the DNA that change the size or charge of a protein molecule. Point mutations that do not cause such changes, and small insertions or deletions, are less likely to be detectable with this method. Currently, null mutations cannot be detected reliably, and this is probably the greatest disadvantage of 2-D PAGE. Overall, 2-D PAGE should detect an estimated 25 to 35 percent of all spontaneous gene mutations. Several aspects of 2-D PAGE of blood proteins, in addition to computerized comparisons of gels, are amenable to improvement, and at least one approach is being developed to improve detection of null mutations (92).

The identity of most of the proteins that can be visualized on two-dimensional gels is unknown. Although it has not yet been done, the technology exists to recover proteins from gels, purify them, and then determine their amino acid sequences (92). Once amino acid sequences are worked out, the nucleotide sequences that code for them can be deduced and nucleic acid probes can be assembled to correspond to the protein-coding nucleotide sequences. The probes could theoretically be used to locate the corresponding DNA within the human genome, thereby determining exactly where in the genome the coding sequence lies.

DNA Sequencing

Conceptually, the most thorough and straightforward analysis of the genome would consist of lining up the chromosomes and determining the nucleotide sequence from end to end on each one. Once the sequence was known, it could be stored in computer memory.

The advantage of determining the complete genomic sequence would be that any nucleotide sequence that was identified in any laboratory could immediately be mapped to its chromosomal location. As nucleotide sequences for various parts of the genome were obtained in different laboratories, both common polymorphisms[1] and rare mutations could be identified by making comparisons between the total genomic sequence and partial sequences. Furthermore, the method would detect mutations and polymorphisms no matter where they occur—in introns, exons, and repetitive sequences.

Walter Gilbert, Fred Sanger, and their colleagues developed the theoretical basis for the techniques that makes it possible to consider sequencing the genome (67,118) and in 1984, George Church and Walter Gilbert described a method for sequencing any part of the genome (20), although the method could be applied to sequencing the entire genome. The genomic DNA would be cut into pieces small enough to be handled in the laboratory, and then chemical methods would be used to determine the nucleotide sequence within each fragment. The methods are well understood, are being applied in dozens of laboratories around the world, and are a standard exercise in the education of molecular biology graduate students (33). Through February

[1]Polymorphisms are "common" mutations that are present in the population at a relatively high frequency. Several alternative versions of the same basic gene sequence, all equally functional, can co-exist. New mutations are defined as occurring in less than 1 percent of the population, excluding the more common polymorphic variants.

1985, the methods have been used to examine discrete and defined segments of DNA from organisms with large genomes and complete genomes from smaller organisms, sequencing a total of 4,045,305 base pairs (150)—roughly 1/800 of the human genome.

Practically, we are far away from being able to sequence the entire genome, not because the techniques are unavailable but because they are expensive and slow when applied to a task of the magnitude of sequencing 3 billion nucleotides. A meeting during the summer of 1985 considered this enormous task and estimated the cost at about $1 per nucleotide, or $3 billion overall. The effort can also be estimated in terms of technician-years, and that comes to about 200 technician-years to sequence 100 million nucleotides, which would be expected to yield about one new mutation.

Although it is conceptually satisfying, the enormous resources necessary to sequence the entire genome make it unlikely that the task will be undertaken soon. However, genomic sequencing technology is still progressing, and further automation of the process may increase its efficiency. Continued encouragement of those efforts and provision of sufficient resources to develop techniques may make it possible to sequence the entire human genome in a reasonable time. When it is done, it will probably be undertaken as part of an effort to provide more general information about human genetics, rather than to identify human mutations. If the genome is sequenced, for whatever reason, it will be a great boon to mutation studies by providing a reference for all techniques that involve sequencing or the use of restriction enzymes.

Restriction Fragment Length Polymorphisms

A method that offers a less exhaustive survey than genomic sequencing, but which is potentially more efficient, involves the use of restriction enzymes and gel electrophoresis to detect nucleotide substitutions or small deletions in the DNA (98).

Genomic DNA is isolated from white blood cells and treated with restriction enzymes, chemicals that recognize specific sequences in the DNA and "cut" the DNA wherever such sequences occur (see fig. 7). (The number of fragments produced is determined by the frequency of occurrence of the particular enzyme recognition site in the DNA sequence.) Restriction site analysis does not, in practice, examine every nucleotide. However, the use of a set of combined restriction enzymes increases the number of restriction sites identified, allowing examination of a larger portion of the DNA, including both expressed and nonexpressed regions.

After the DNA is fragmented (see fig. 8), there are two alternative ways of proceeding, one using a gene cloning method, and the other using a noncloning, direct method.

Cloning is the process of "growing" human DNA in other organisms; it reduces the amount of DNA needed—in most cases corresponding to the amount of blood that must be drawn—but it increases the number of laboratory manipulations. Additionally, it introduces some uncertainty about the possibility that DNA changes will occur during cloning. This can be checked, but requires diligence.

In the gene cloning method, the human DNA fragments are incorporated into the genetic material of a virus called "lambda." Lambda reproduces in the bacterium *E.coli*, and large quantities of viral DNA containing the human segments are produced. Following isolation of the cloned human DNA, each clone is treated with a set of restriction enzymes that cut it into smaller fragments. These collections of fragments are then separated by gel electrophoresis, producing distinct bands visible by gel staining methods. A visual comparison of the location of the bands derived from parental DNA and from children's DNA shows whether a mutation has occurred in a site for a restriction enzyme. The presence of a band in the child's DNA that is not present in the parents' DNA, or vice versa, indicates a nucleotide substitution or deletion restriction enzyme recognition site and suggests that a heritable mutation has occurred.

Figure 7.—Production of Restriction Fragment Length Polymorphisms Using Restriction Enzymes and Separation of Different DNA Fragment Sizes by Agarose Gel Electrophoresis

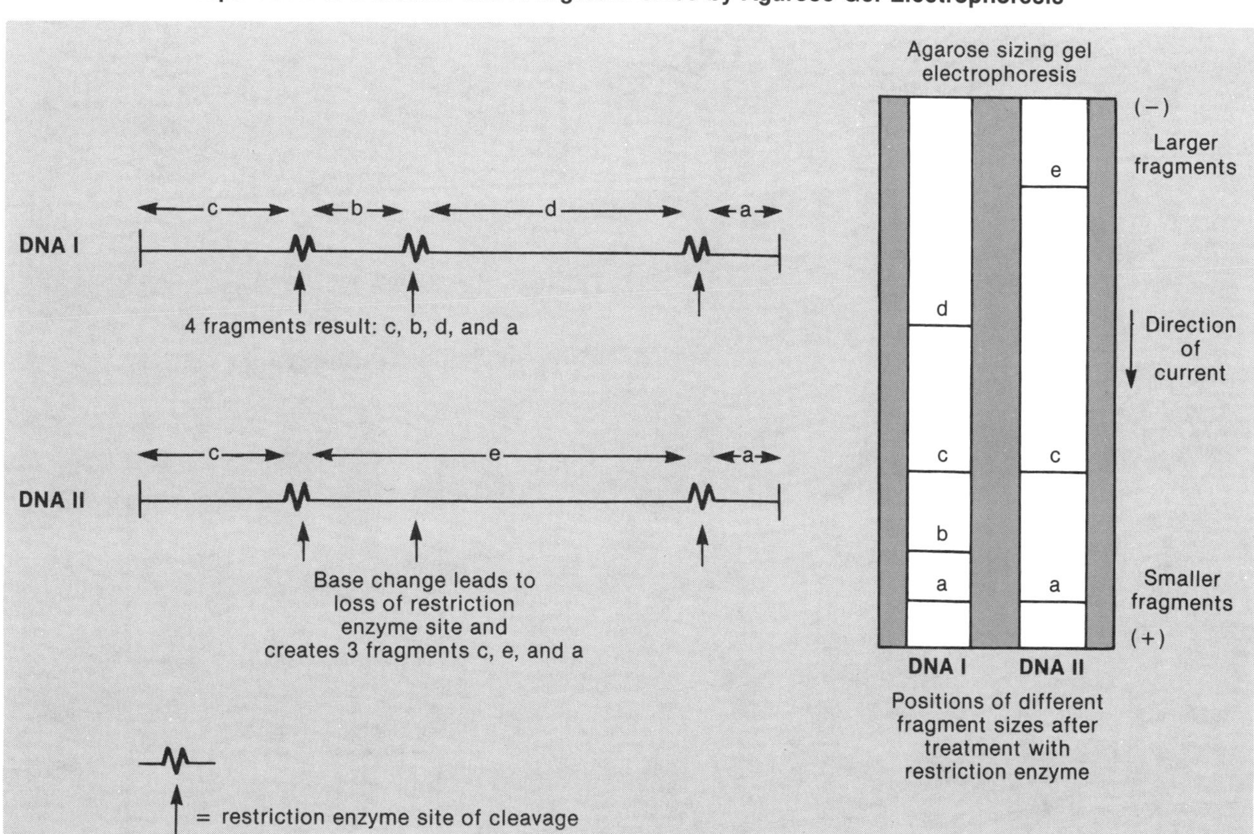

SOURCE: Office of Technology Assessment.

An alternative method bypasses the lambda cloning procedure, and analyzes DNA fragments directly. Fragments from the entire human genome are spread out on an electrophoretic gel, forming smears of indistinguishable bands. In order to visualize the location of particular segments of DNA and to evaluate their position relative to the equivalent band in the reference sample, the naturally double-stranded DNA is dissociated into single-stranded DNA while still in the gel and allowed to reassociate with small pieces of specific genes which are added to the gel and which are radioactive (probes). Those original DNA sequences that pair with the probes are now tagged. The DNA is then transferred to a flexible plastic membrane conserving the spatial arrangement of the DNA (a procedure known as Southern blotting) and autoradiographed; bands are detected where the tagged sequences are present. As with the cloning method above, the appearance of a new band or the disappearance of an old band suggests a nucleotide substitution or deletion in a recognition site of one of the restriction enzymes.

One-Dimensional Denaturing Gradient Gel Electrophoresis

A modification of the standard electrophoretic gel procedure, proposed by Fischer and Lerman (34), allows DNA to be separated not only on the basis of size, but also on the basis of sequence of nucleotides even if differences do not occur at restriction enzyme recognition sites. Double-stranded DNA dissociates into single-stranded DNA when it is heated or when it is exposed to denaturing chemicals (e.g., formamide or urea). A gradient of increasing strength of such chemicals can be produced in a gel so that DNA samples will travel in the direction of the electric cur-

Figure 8.—Restriction Fragment Length Polymorphism

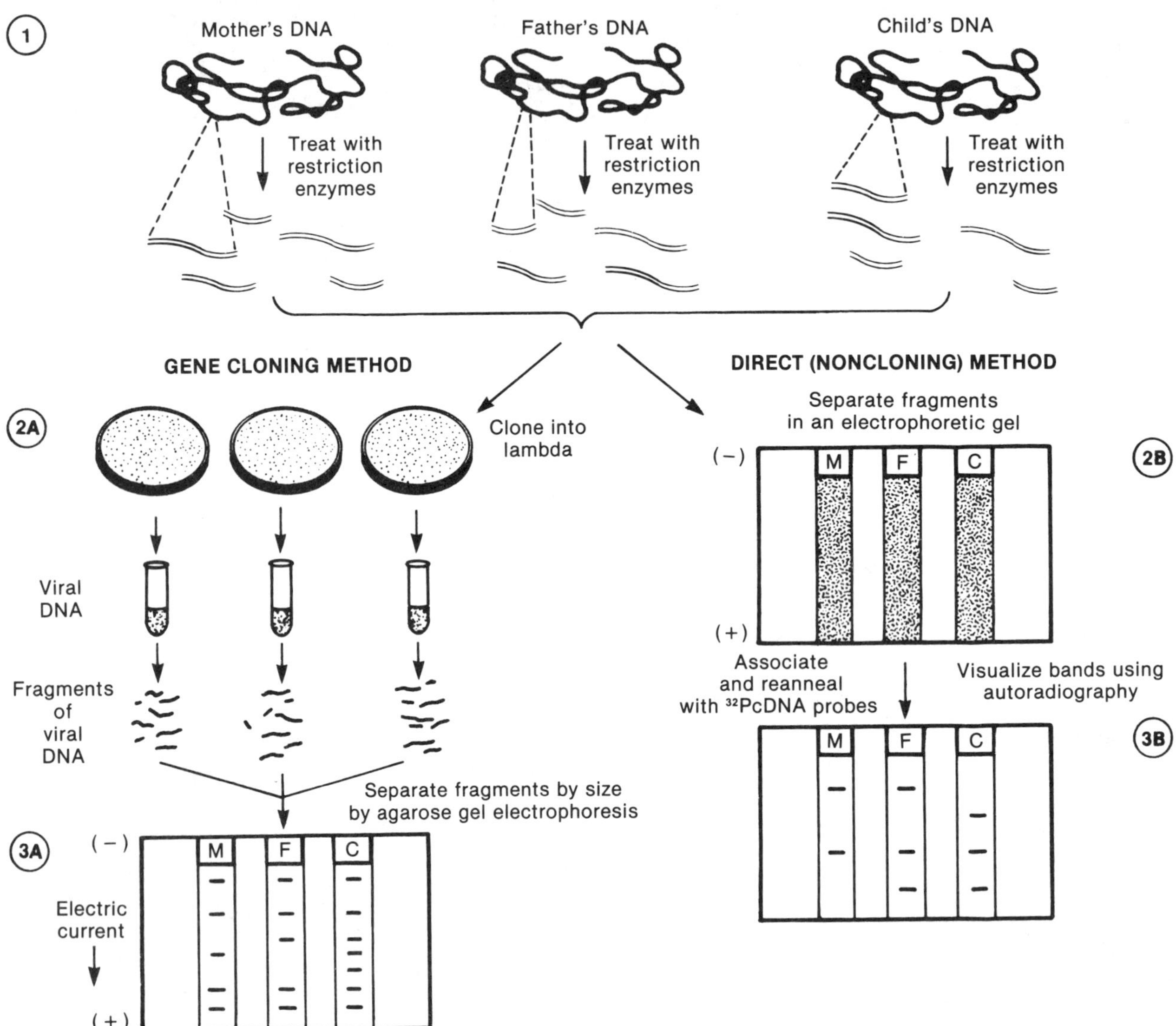

1 Isolate geonomic DNA from white blood cells. Using restriction enzymes, cut DNA into double-stranded fragments of various lengths.

Gene Cloning Method:

2A Clone each fragment into the bacterial virus lambda, infect into *E.coli*, grown in petri dishes, and allow the virus to replicate within the bacteria. Isolate large quantities of the viral DNA, which now contains segments of human DNA, and cleave it with restriction enzymes.

3A Separate DNA fragments using agarose gel electrophoresis. Visualize bands corresponding to different sizes of fragments using fluorescent staining. Fragments found in the child's DNA samples and not in the parents' DNA may contain heritable mutations. These bands can be removed and the DNA analyzed for specific mutations.

Direct (noncloning) Method:

2B Apply samples of DNA fragments to an electrophoretic gel, producing smears of indistinguishable bands.

3B Dissociate DNA fragments into single strands and incubate with radioactive, single-stranded, ^{32}P-labeled DNA probes for specific human genes. The probes hybridize with complementary sequences in the DNA and label their position in the gel. Visualize the position of bands using autoradiography. Changes in the position of bands in the child's DNA compared to the parents' DNA suggest possible new mutations. Bands can be isolated and DNA analyzed for sequence differences.

SOURCE: Office of Technology Assessment.

rent, separating by size, and will also begin to dissociate as they reach their particular critical concentration of denaturing chemical. Dissociation causes the molecule to split into constituent parts, or unravel, and get stuck in the pores of the gel (84).

Every unique strand of DNA dissociates at a unique concentration of denaturant. In fact, the difference of only one nucleotide between two otherwise identical strands of double-stranded DNA of 250 nucleotides in length is enough to cause the strands to dissociate at different concentrations of denaturant chemical, and to stop travelling at different locations in the gel. Again, a comparison between the banding pattern of parents' and child's DNA analyzed in this way may identify a wide range of mutations in all regions of the DNA.

In this method, total genomic DNA is isolated from individuals and is heated to form single-stranded DNA (see fig. 9). It is then mixed with radioactive, single-stranded DNA probes that correspond to a small portion of the genome. Following incubation so that double-stranded hybrid molecules ("heteroduplexes") will form, they are treated with restriction enzymes and electrophoresed in a denaturing gradient gel. The presence of a single mismatch between a nucleotide in the "normal" probe and the corresponding segment from a person bearing a mutation makes the heteroduplex slightly more sensitive to denaturing chemicals so that such a molecule will stop traveling in the gel before the point at which a perfectly paired heteroduplex will stop. The gel is dried and exposed to X-ray film, and the resulting autoradiogram is examined visually for differences in banding patterns between parents' and child's DNA.

Two-Dimensional Denaturing Gradient Gel Electrophoresis

Leonard Lerman has proposed a technique whereby a two-dimensional separation of parents' and child's DNA is used to differentiate among DNA sequences common to all three members, polymorphisms in either parent which are transmitted to the child, and new mutations in the child's DNA (57). Like the two-dimensional polyacrylamide gel procedure for protein separation described above, two-dimensional denaturing gradient gel electrophoresis (2DGGE) compares locations of spots on a gel (in this case, DNA spots) representing various types of mutations in the nucleotide sequence in expressed and nonexpressed regions of the DNA. This method separates DNA fragments on the basis of differences in base composition (or sequence), after separating them on the basis of differences in length. 2DGGE detects differences between DNA heteroduplexes by responses of their structure to gradual changes in denaturant concentration in the gel.

Genomic DNA from the parents in one sample and from the parents and children in another sample would first be treated with restriction enzymes and separated by size on separate agarose gels in a single lane (see fig. 10). The gel strips are then laid across the top of two denaturant gradient gels and the DNA is electrophoresed through the increasing gradient of denaturant. Each gel is then cut into horizontal slices, each of which contains pieces of the original homoduplex DNA of all sizes, but including only those pieces that dissociated at the same denaturing concentration ("isomelting" groups). These gel slices containing homoduplexes of parents' DNA or of parents' and child's DNA are physically removed (or collected on removable membranes) heated, and allowed to reassociate as heteroduplex molecules (see fig. 10).

In order to differentiate between common areas of the DNA among parents and children, another gradient denaturing gel is used to denature the mixtures of double-stranded, isomelting DNA of the parents ($M_1/M_2 + F_1/F_2$) and another of the parents' DNA mixed with the child's ($M_1/M_2 + F_1/F_2 + C_1/C_2$). Various double stranded combinations of parents' and child's DNA are produced (e.g., M_2/C_1, F_1/C_2, etc.). The two sets of mixtures are compared in the final two gels.

Sequences that pair up perfectly represent common sequences to all three (level "i" in fig. 10). At the bottom, a narrow region, dense with heteroduplex fragments, represents the largest fraction of the sample—those that are perfectly matched between base pairs and therefore are located at the same denaturant concentration as the original homoduplexes contained in the narrow slice

Figure 9.—One-Dimensional Denaturing Gradient Gel Electrophoresis

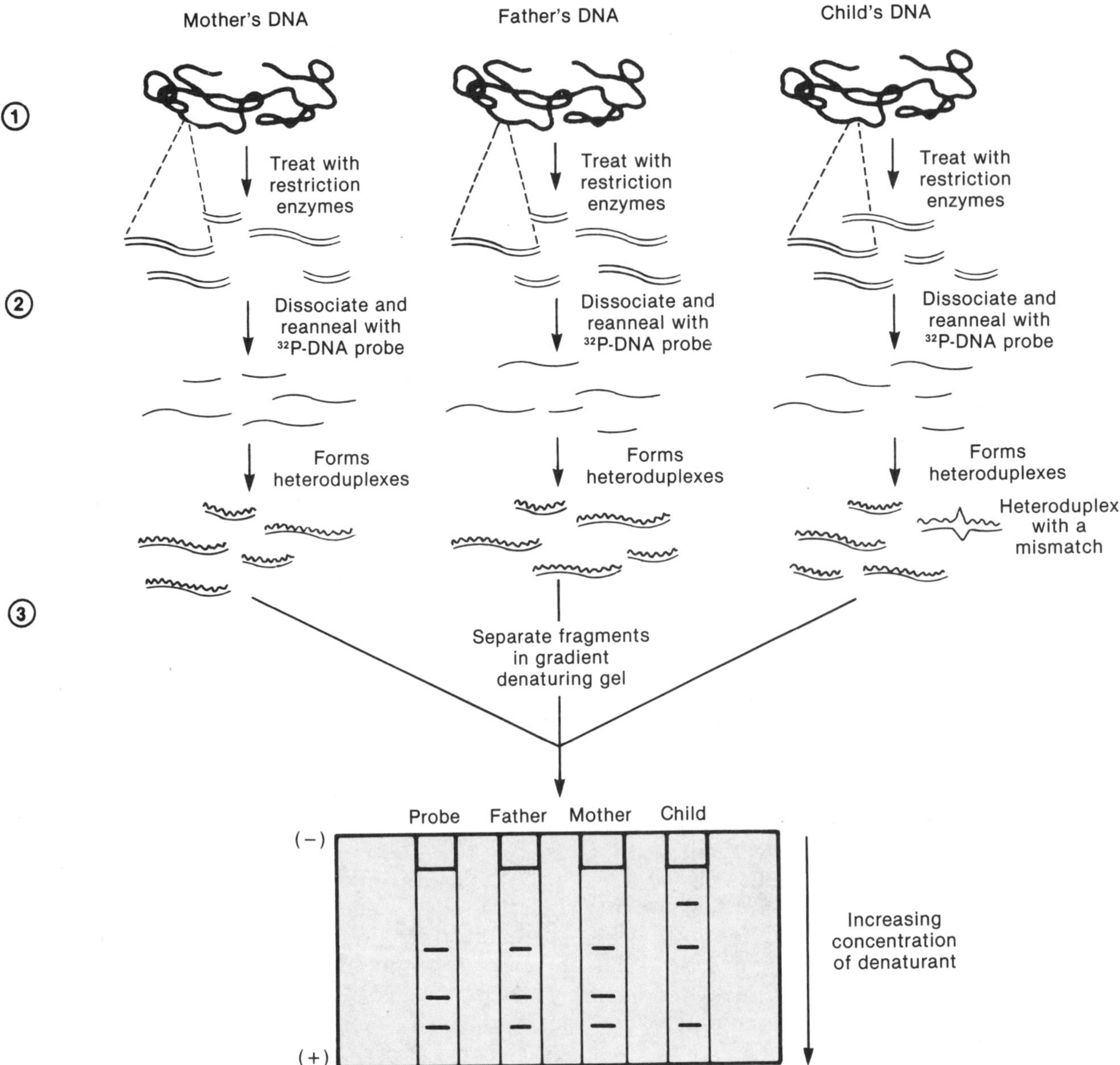

(1) Isolate genomic DNA from white blood cells. Using restriction enzymes, cut DNA into (double-stranded) fragments of various lengths.

(2) Dissociate double-stranded DNA fragments into single strands, and reanneal in the presence of radioactive ^{32}P-labeled, single-stranded DNA probes. Heteroduplexes form between probe and sample DNA even if the base sequences are not perfectly complementary; if mutations are present, some of these heteroduplexes will contain mismatches.

(3) Separate heteroduplex fragments in a denaturing gradient gel. Fragments with mismatches denature in a lower concentration of denaturant than fragments that are perfectly complementary. Visualize the position of the fragments with autoradiography. Fragments containing the child's DNA that denature sooner than the parents' fragments can be analyzed for new mutations.

SOURCE: Office of Technology Assessment.

Figure 10.—Two-Dimensional Denaturing Gradient Gel Electrophoresis

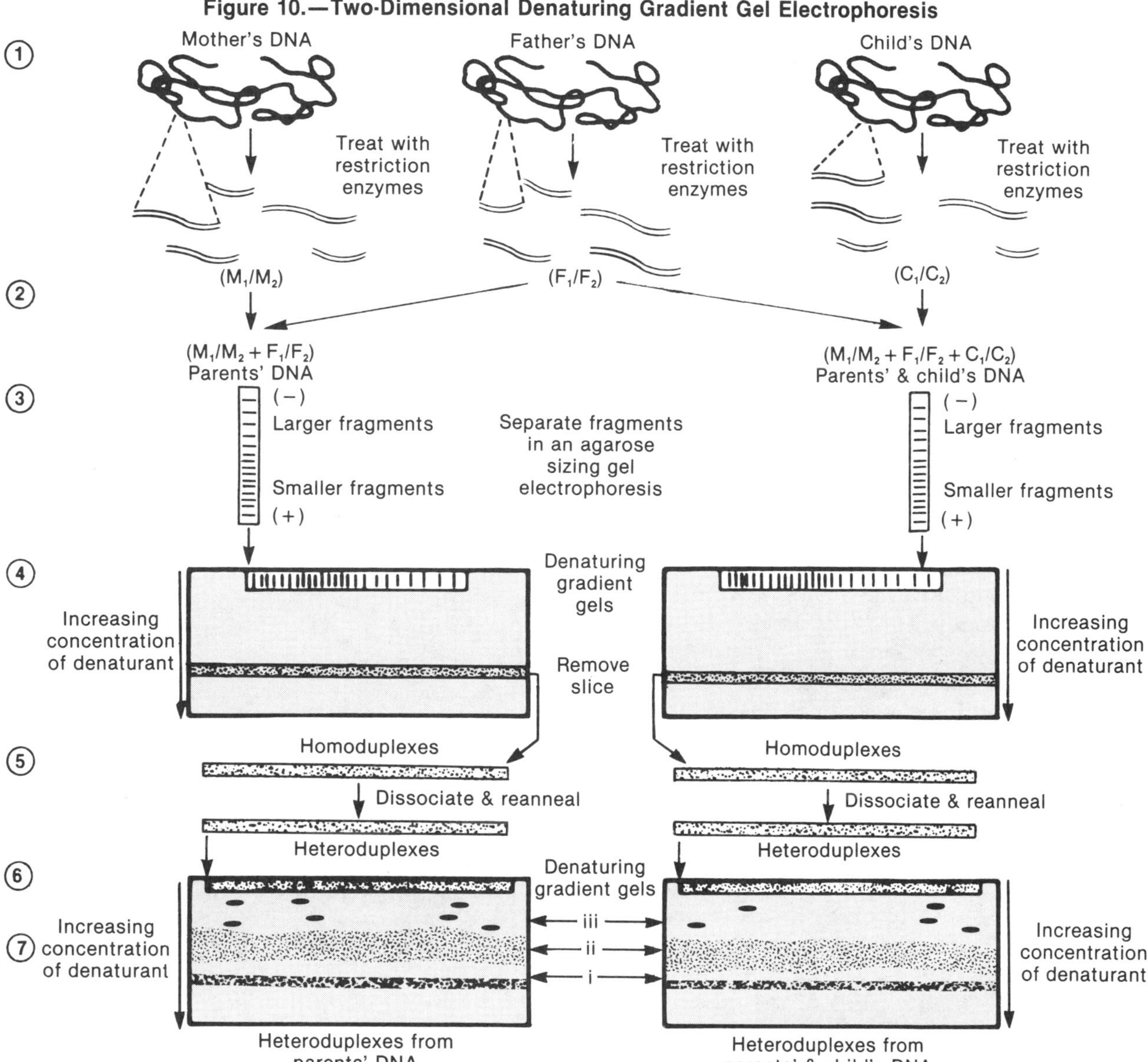

(1) Isolate genomic DNA from white blood cells. Using restriction enzymes, cut DNA into double-stranded fragments of various lengths.

(2) Mix mother's and father's DNA fragments ($M_1/M_2 + F_1/F_2$) in one set, and mix parents' and child's DNA fragments ($M_1/M_2 + F_1/F_2 + C_1/C_2$) in another set.

(3) Separate fragments by length in an agarose sizing gel. Cut out a lengthwise strip of the gel, containing a range of fragment sizes, and lay the strip across the top of a denaturing gradient gel.

(4) Separate fragments along an increasing concentration of denaturant. Cut out a slice of the gel at a narrow interval of the temperature gradient, or collect a similar interval on a removable membrane.

(5) While maintaining spatial arrangement of fragments in the narrow slice, dissociate the double-stranded fragments into single strands, and reanneal them to form heteroduplexes (e.g., M_2/C_1, F_1/C_2, F_2/M_1, etc.) of various combinations of strands of the original homoduplexes (M_1/M_2, F_1/F_2, and C_1/C_2).

(6) Lay the thin strip of heteroduplex fragments arranged by size across the top of a new gradient gel, and separate in the direction of an increasing concentration of denaturants.

(7) Comparing the two final gels, the DNA spots are grouped in three distinct regions (see text for explanation). Analyze spots in the highest region of the gel for DNA fragments with possible new mutations.

SOURCE: Office of Technology Assessment.

removed from the gel or collected in the membrane.

Polymorphisms that are present in one parent or the other show up as spots that denature earlier than the perfectly matched DNAs (level "ii"). In the middle region of the gel, a diffuse distribution of fragments represents heteroduplexes with single mismatches between the two strands, causing them to dissociate earlier in the gel (at a lower denaturant concentration, occurring above the perfectly matched heteroduplexes). This region would contain inherited polymorphisms as well as new mutations, but the region would be too dense with DNA spots to distinguish individual spots.

Molecules that contain more than one polymorphism per fragment denature even earlier (level "iii") since multiple mismatches in a fragment have an additive effect on the fragments stability in the denaturing gradient. The highest area in the gel, where heteroduplexes dissociate in the lowest denaturant concentration, is where fragments with more than one mismatch would be located. These mismatches could be new mutations or inherited polymorphisms. Since the number of such fragments in the sample is likely to be small, this region would show discrete, nonoverlapping spots; DNA spots in this third level of the gel can be removed from the gel and analyzed to identify possible new heritable mutations. By comparing the two final gels, new spots should be apparent to a trained eye. These would represent single base pair changes or very small deletions and rearrangements in all regions of the DNA.

In theory, it should be possible to cut many strips of isomelting regions out of the first gel and derive an analysis of each melting region by a series of second gels. By comparing the child's DNA to his or her parents' DNA, this approach allows for large portions of the genome to be examined simultaneously.

Ribonuclease Cleavage of Mismatches in RNA/DNA Heteroduplexes

A technically simpler approach to detecting mutations in cloned and genomic DNA has been proposed by Richard Myers (85,86). This technique uses an enzyme, ribonuclease A (RNaseA), that cleaves double-stranded RNA/ DNA heteroduplexes where a specific mismatch of base pairs occurs. RNaseA cleaves the RNA/ DNA molecule where a cytosine (C) in RNA occurs opposite to adenine (A) in DNA. (C normally pairs with G [guanine] and A normally pairs with T [thymidine].) The idea behind this method is similar to that of restriction enzymes, each of which cleaves DNA at specific normal sequences, except that RNaseA cleaves RNA/DNA hybrid molecules where mismatches occur. The efficiency of this approach depends on the number of different mismatches that can be recognized and cleaved. The greater the number of enzymes that can be found to cleave different mismatches, the more types of mutations can be detected. This method should detect nucleotide substitutions over a large portion of the DNA.

With this method, genomic DNA is first isolated and mixed with radioactively labeled RNA probes (see fig. 11). The mixture is heated so that the strands dissociate and then reassociate randomly with complementary strands. When RNA molecules bind with DNA molecules of homologous sequences, the result is one type of "heteroduplex." "Homologous" means that the sequence is close enough between the two molecules that they will form a stable double-stranded molecule. The presence of a single base pair mismatch does not prevent double-stranded heteroduplexes from forming, however.

The heteroduplexes are treated with RNaseA and separated according to size on a standard agarose gel. Perfectly paired molecules will be unaffected by RNaseA and will form bands on the gel that correspond to their original size. However, imperfectly paired molecules will be cut by the enzyme at the site of the C:A mismatch, and two separate fragments will result, showing up as two separate bands on the gel. For analysis, parents' and child's DNA are treated and electrophoresed separately, and the gels compared for different patterns by autoradiography to reveal the occurrence of any possible mismatches. Currently this method is applicable only for (RNA)C: A(DNA) mismatches, which represent 1 of 12 possible types of mismatches between these heteroduplexes.

Figure 11.—Ribonuclease Cleavage of Mismatches in RNA/DNA Heteroduplexes

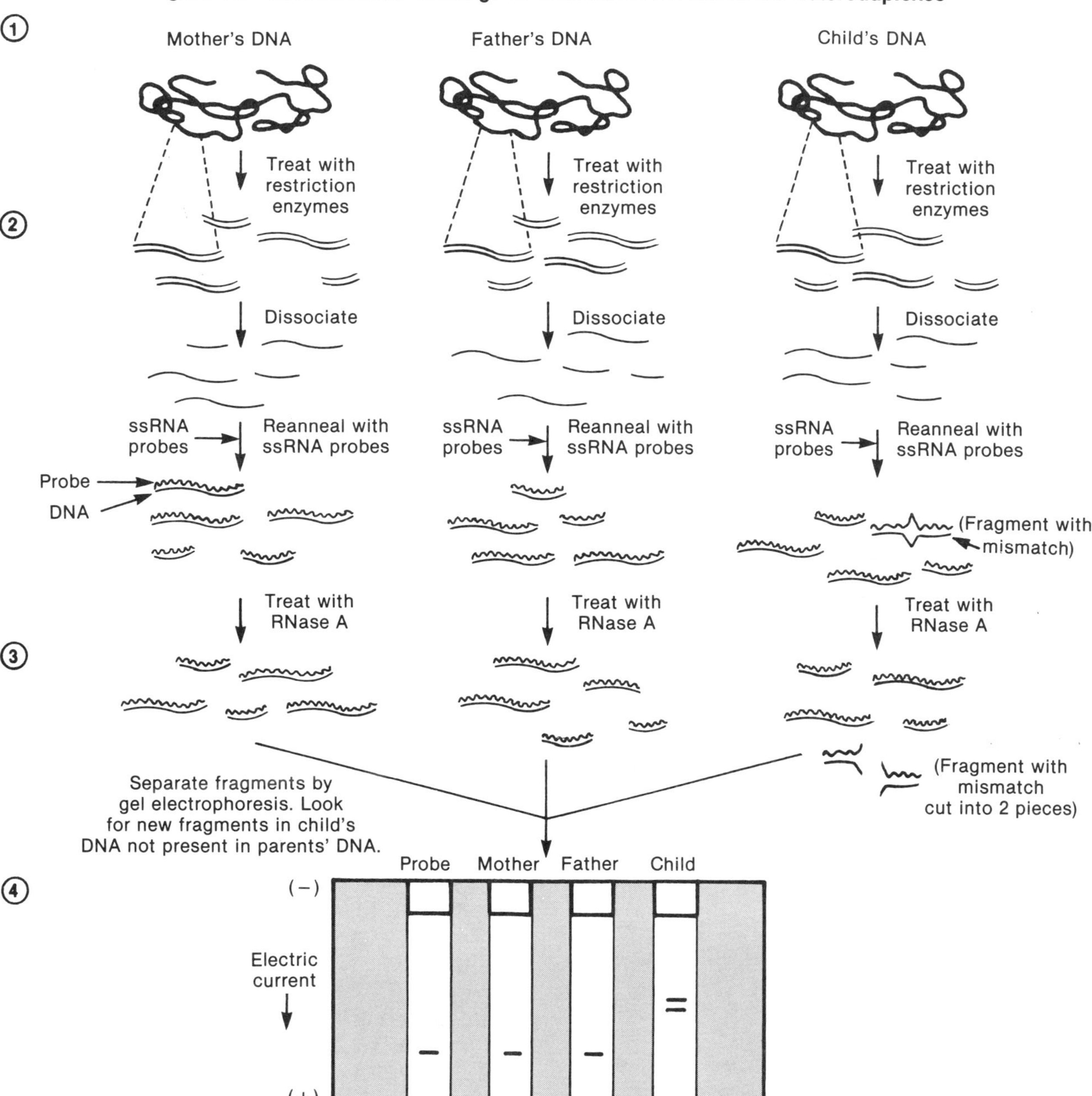

(1) Isolate genomic DNA from white blood cells. Using restriction enzymes, cut DNA into double-stranded fragments of various lengths.

(2) Dissociate double-stranded DNA fragments to single strands, and reanneal in the presence of single-stranded RNA (ssRNA) probes, forming RNA/DNA heteroduplexes.

(3) Treat the heteroduplexes with RNaseA, an enzyme that cuts RNA/DNA heteroduplexes where certain mismatches occur (e.g., where cytosine in RNA is mismatched with adenine in DNA).

(4) Separate heteroduplex fragments by size using gel electrophoresis. Visualize the position of fragments in the gel by autoradiography. Fragments found in the heteroduplexes containing the child's DNA, but not found in those containing parents' DNA, indicate possible new mutations.

SOURCE: Office of Technology Assessment.

Subtractive Hybridization

Detecting mutations would be much easier if it were possible to ignore the millions of nucleotide sequences that are identical in both parents and child and, instead, focus only on the relatively few sequences that are different. Currently, the best estimate for the frequency of human mutations is about 1 in 100 million base pairs, or about 30 per genome, so it is necessary to examine 3 billion base pairs from each of the parents and a child and find those 30 that differ. George Church had conceived of a method to find sequences in a child's DNA that are not present in either parent's DNA (25).

While there is no experience with Church's proposed method, and therefore, no information about its feasibility, it is a promising idea. The basis of Church's method is to view the human genome as a group of unique nucleotide sequences. In this context, "unique sequence" means a sequence that may occur only once in the entire genome. To explain what a unique sequence is, consider the shortest and longest sequence in the genome, a single nucleotide and the entire genome. A single nucleotide is not unique; A, T, G, and C occur repeatedly throughout the genome. At the other extreme, the entire genome is unique. Somewhere in between those extremes is a length of nucleotides that is long enough to be unique without being too long to handle experimentally.

It turns out that a sequence of 18 nucleotides is long enough to be unique; on statistical grounds, any sequence of that length or greater that occurs once in the genome is not expected to occur a second time. Therefore, if every possible sequence of 18 nucleotides, or "18-mer," is synthesized, this collection of 18-mers will include sequences that together represent every possible sequence of 18 nucleotides of DNA (there are 4^{18} or 70 billion possibilities). Some proportion of these are actually present in human sequences, some are not. It is feasible to make the entire collection of 70 billion different 18-mers, and a single synthesis provides sufficient 18-mers for many experiments.

To look for mutations, DNA from parents and child is isolated and cut with restriction enzymes into relatively small lengths of 40 to 200 base pairs and dissociated into single strands (see fig. 12). This parental DNA alone is mixed with the collection of 18-mers under conditions that permit the formation of perfect hybrids: each nucleotide sequence of an 18-mer hybridizes exactly with each corresponding genomic sequence. The conditions are also such that an 18-mer that interacts with a genomic sequence complementary to just 17 of the 18 nucleotides will not form a stable hybrid. When the hybridization reaction is complete, the 18-mers that did not find a perfect match with any parental DNA sequences are left behind as single-stranded molecules, unchanged by their participation in the hybridization reaction. Because they are single stranded, they can be separated from all the other 18-mers that are hybridized to parental DNA. This unbound fraction of the mixture is retained, while the fragments of parental DNA that bind to the 18-mers are discarded.

This pool of single-stranded 18-mers which did not bind to any parental DNA sequence is then mixed with similarly prepared DNA from the child, and hybridized under the same conditions. This time, however, the molecules of interest are the sequences that bind to the child's DNA (and not previously to the parents' DNA). Any 18-mer that hybridizes perfectly with a sequence in the child's DNA will identify a sequence that is present in the child's DNA and absent from the parents' DNA. Such sequences may contain new mutations present in the child's DNA. It should then be possible to isolate and characterize this hybrid formed between the 18-mer and child's DNA, so that the mutant sequence can be determined. If mutations occur at a frequency of 1 per 100 million nucleotides, and if this technique works perfectly, it could detect 30 to 40 nucleotide sequences in the child's DNA that are not present in the parents' DNA.

This approach is the least well developed of all the ones discussed in this report, and its feasibility is unknown. If it does prove feasible, this approach could identify short sequences containing mutations in any part of the DNA, allowing further detailed study (e.g., by DNA sequencing) of the kinds of mutations that are occurring.

Figure 12.—Subtractive Hybridization

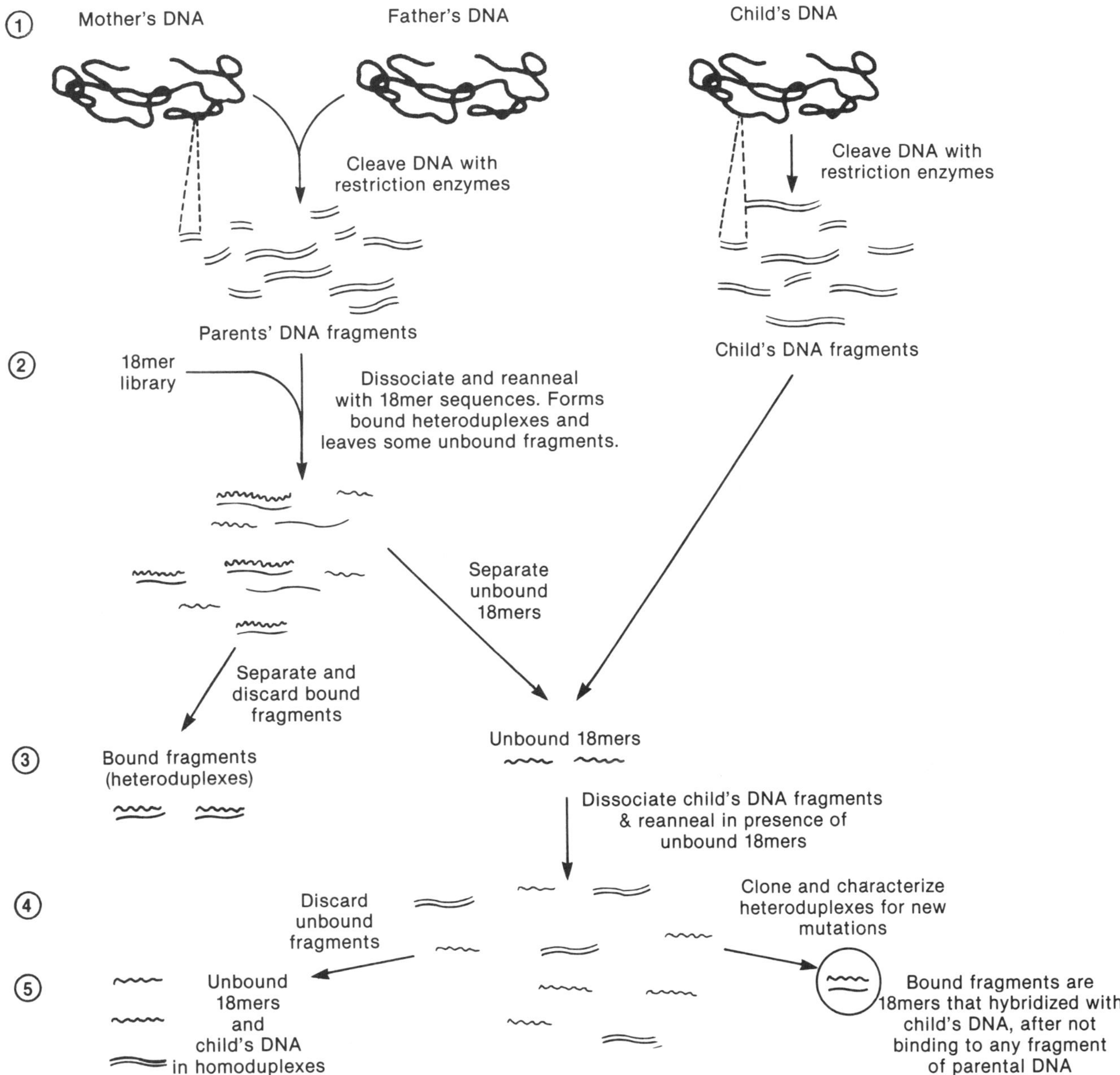

(1) Isolate genomic DNA from parents' and child's white blood cells; mix parents' DNA samples together, while keeping child's DNA separate. Cleave DNA into fragments with restriction enzymes.

(2) Dissociate parents' DNA fragments into single strands, and reanneal with a set of unique single-stranded 18mer sequences. 18mer sequences complementary to sequences in the parents' DNA will form double-stranded, or bound, fragments. The remainder, the unbound fraction, will remain single stranded.

(3) Separate the bound fraction from the unbound fraction. The bound fraction is discarded.

(4) Unbound 18mers for which no parental complement exists are allowed to hybridize with the child's DNA fragments under the same conditions of dissociation and reannealing. Any fragments of the child's DNA that hybridize with unbound fraction of 18mers represent sequences present in the child's DNA that are not present in either parents' DNA.

(5) Isolate these heteroduplexes and analyze them for specific new mutations.

SOURCE: Office of Technology Assessment.

Pulsed Field Gel Electrophoresis

If human DNA were short and simple, it could be cut up with restriction enzymes and separated electrophoretically into discrete bands, each representing a particular segment of the total DNA. However, human DNA is so long that electrophoretically separated fragments make a continuous smear of bands. If the DNA were cut in only a few places to produce only 100 or 200 bands, the pieces would be too big to pass individually through the pores of a standard electrophoretic gel. A new technique, pulsed field gel electrophoresis (PFGE) is being developed to allow separation of fragments of human DNA larger than can be separated with other electrophoretic techniques and to examine such fragments for evidence of mutations. In theory, the procedure will detect submicroscopic chromosome mutations, including rearrangements, deletions, breaks, and transpositions. At present, the method cannot handle intact human chromosomes although it works very well with smaller chromosomes from lower organisms and with small fragments of human chromosomes (157,158).

The standard process for isolating genomic DNA randomly breaks the long molecules into pieces. In the PFGE procedure, genomic DNA is treated so that random breaks are avoided. Whole cells are suspended in liquid agarose and after the agarose solidifies into a gel, enzymes are added to degrade proteins and RNA. This apparently minimizes random damage to the high molecular weight DNA. DNA is cut into exact and predictable fragments by addition of a "rare cutter," a restriction enzyme (or group of enzymes) that cuts DNA consisting of 6 billion nucleotides, into only 3,000 pieces averaging 2 million nucleotides each (about 1/60 of an average chromosome) (126).

Ten different fragments are each labeled with a radioactive probe, so that each lane in the gel will have 10 visible bands, the most that can be analyzed per lane. The fragments are then separated on an agarose gel in which the electrical field is applied alternately ("pulsed") in perpendicular directions for 24 to 72 hours. The pulse time is adjusted, according to the particular size of the fragments, to maximize separation of the fragments. The optimal pulse time is one in which the fragments are constantly untangling and reorienting themselves. Autoradiography is the final step in the procedure. Each of the 10 fragments bound to a labeled probe appears as a visible band on the gel on exposure to X-rays; chromosomal mutations would appear as a shift in the position of the fragment containing the mutation.

Recent evidence suggests that fairly large pieces of human chromosomes can be separated, although most of the experience with the technique is limited to certain lower organisms such as yeasts and unicellular parasites. Results of experiments with DNA derived from these organisms, which have much smaller genomes than humans, suggest that chromosome mutations would be apparent if they were at least 5 percent as large as the fragment itself. The procedure would have to be modified for human chromosomes due to their larger size and complexity (127). This technique may be useful in detecting chromosome mutations that are intermediate in size between major rearrangements (observable by cytogenetic methods) and single base pair changes.

CONCLUSIONS

In a short period of time, enormous advances have been made in the ability to "read" the genetic material and to understand what it means. The techniques described in this chapter are examples of state-of-the-art molecular genetics. Some of the components of these techniques were originally developed for other purposes in genetic research, while others were designed specifically for detecting new heritable mutations. They all are successes in the sense that they propose reasonable and verifiable ways of examining human DNA for alterations in the nucleotide sequence. However, none of the DNA-based techniques is approaching the efficiency that would be needed for any of them to be appropriate for field use. As these technologies develop and information is shared among investigators, it may be possible to design combinations of methods that complement each other and that select for those samples in which there is a greater likelihood of finding new mutations.

In general, despite impressive achievements that have been made, the technology remains too unwieldy for large-scale applications. Olson (98) likens the current situation in mutation research to the status of computer technology in the late 1950s. The spectacular improvement in computers came not from scale-up of 1950s technology but from development of the integrated circuit. He suggests that big improvements in technology will provide the appropriate methods; doing more of what we have already done is unlikely to be sufficient.

Analyzing complete sets of human DNA from large numbers of people using these methods would, at present, require an enormous commitment of resources. In considering the wisdom of devoting major resources to such research, Olson (98) emphasizes that the program's ultimate goal would have to be far more ambitious than the detection of a few bona-fide examples of new mutations. A program large enough to have a legitimate impact on public policy about environmental exposures would have to be much larger than a 5 or 10 person laboratory and be several orders of magnitude more efficient than existing restriction fragment length polymorphism technology.

One of the outstanding characteristics of current research in molecular genetics is the rapid pace of new developments and ideas; assuming this continues or accelerates, we can count on better methods to replace the proposed ones described in this report. The better ones should be able to examine at least the major portions of the genome and be practical in studying populations at higher risk for heritable mutations. They should also be able to detect a wide, if not the entire, spectrum of mutational events.

Until the next generation of techniques is developed, the best course for the next few years may be to maintain a balance between the support of current methods and the support of basic research which is the source of better ones. Maintaining a balance implies that resource-intensive efforts should not pull unwarranted amounts of resources away from developing new techniques, although no special scientific effort seems to be needed to encourage this highly innovative and rapidly paced area of research.

Chapter 5

New Methods for Measuring Somatic Mutations

Contents

TABLE

Chapter 5

New Methods for Measuring Somatic Mutations

INTRODUCTION

All the techniques for learning about heritable mutations described in chapter 4 rely on comparing DNA or proteins of parents with DNA or proteins of their children to infer the kinds and rates of mutations that occurred in parental germ cells. While that information is of great value in learning about heritable mutations, it comes well after the mutations have actually occurred. Heritable mutation tests are not useful for determining rapidly that an individual or a population has been exposed to a mutagen, so that people who have been exposed can be appropriately monitored and others, who have not been exposed, can be warned and spared exposure to that same mutagen. For this reason, it is desirable to develop tests for *somatic cell* mutations that might allow inferences to be made about possible heritable effects of some mutagenic exposure. Instead of providing an exhaustive catalog of somatic cell mutation assays, this chapter provides examples of developing technologies that may soon be available to measure somatic mutations in human populations.

It is not known whether the rates and kinds of somatic mutations in blood or any other nongerminal tissue have any predictable relationship with rates and kinds of germinal mutations; it is only with further development and testing of those methods, together with methods for detecting germinal mutations, that such relationships might be elucidated. However, even without knowing the precise relationships between somatic and germinal mutations, somatic tests may provide an early warning that people are being exposed to environmental mutagens. Furthermore, identifying and characterizing somatic mutants will provide important guidance for reducing exposures to mutagens and carcinogens, and is relevant to the study of aging and of carcinogenesis.

Exposure to mutagens may increase the risk of developing some kinds of cancer. In addition, increases in the frequencies of second cancers occurring years after treatment of a first cancer with radiation therapy have been noted (see e.g., ref. 11). In a study of survivors of cervical cancer, the risk of developing a second cancer was found to be highest for organs close to where radiation therapy was delivered, strongly suggesting that the development of second cancers may be related to the mutagenic effects of radiation (11).

Measurement of somatic mutants offers the possibility of identifying differences in somatic mutation rates over time within individuals, among different individuals, and among different populations. Such studies can be used to examine possible associations between exposures to particular environmental or occupational agents and the frequency of mutations. Ultimately, they may be used to identify individuals or populations who are exposed to mutagens. This information could be useful to the regulatory agencies in identifying genotoxic compounds and their potencies.

To be useful for studying human populations, somatic tests must be based on reasonably accessible cells (e.g., blood cells) and on recognizable markers for mutation (e.g., variant proteins or enzymes). The most easily sampled human tissue is blood, and thus far, all the somatic tests that have been developed for studying mutations have used red and white blood cells. Somatic cell mutation assays focus on two types of cellular changes that result from mutations: 1) alterations in phenotypic properties of cells, such as changes in cells' resistance to drugs added to their growth media;[1] and 2) alteration in a gene product, such

[1] Studying drug resistance in humans is inherently limited to studying single copy genes, such as genes on the X chromosome (including the *hprt* gene) or dominant selectable markers.

as a protein. The first type of change has been most extensively studied using the mutation assay based on 6-thioguanine (6-TG) resistance, and the second type has been studied using two types of proteins found in circulating red blood cells.

Instead of examining DNA directly, current and developing somatic tests detect rare mutant cells based on marker phenotypes among large numbers of mostly identical normal cells of a specific type. Although somatic cell mutation assays detect phenotypic events that represent, as closely as possible, actual mutations in DNA, the data are expressed as numbers of *mutants*, rather than in numbers of *mutations*. Various analytic techniques have been used to estimate the mutation frequency from the observed mutant frequency (97).

DETECTION OF SOMATIC MUTATIONS IN WHITE BLOOD CELLS

Albertini and his colleagues (3) and Morley and his colleagues (79) have developed assays using human peripheral blood lymphocytes to detect mutations that result in a phenotypic change in these cells. The assays are based on the ability of mutant T-lymphocytes, which are present in the peripheral blood at the time the sample is drawn, to replicate DNA and to grow under conditions that do not allow growth of normal cells. The mutant cells are identified by their resistance to 6-TG, a drug that is normally toxic to human T-lymphocytes.

Normal cells produce an enzyme, hypoxanthine-guanine phosphoribosyl transferase (HPRT), that metabolizes 6-TG to a toxic chemical, killing the cells. Mutations in the *hprt* gene (the gene that directs the production of HPRT), can result in cells that do not produce HPRT. The mutation from the normal gene ($hprt^{+}$) to the mutant gene ($hprt^{-}$) makes the cell resistant to 6-TG (6-TG^{r}) because they cannot metabolize 6-TG into its toxic form.

This particular selective system has been shown to meet all the necessary criteria for a valid genetic assay: 1) the 6-TG^{r} mutants are genetic variants that "breed true": whether grown with or without 6-TG, they remain resistant to 6-TG when challenged with the drug (3,79); 2) exposure of laboratory cultures of T-lymphocytes to known mutagenic agents such as X-rays or ultraviolet radiation increases the frequency of 6-TG^{r} mutants, indicating an increase in mutation in these cells (17,160); 3) there is a demonstrable change in the gene product—a deficiency in the enzyme HPRT; and 4) there is a demonstrable change in the DNA sequence of the *hprt* gene—a direct demonstration that the *hprt* gene has mutated (4,143).

Selection for HPRT Mutants in T-Lymphocytes

Nicklas (97) described two methods for identifying mutant T-lymphocytes in human blood samples. The first is the autoradiographic assay. After blood is drawn, the T-lymphocytes are purified and frozen. On thawing, the cells are split into two cultures, both of which are treated with phytohemagglutinin, a chemical that stimulates DNA synthesis. One of the cultures is also treated with 6-TG. The other culture is an untreated control.

A label, radioactive thymidine, is added to both cultures, and all cells capable of DNA replication and growth incorporate the labeled thymidine into their DNA. In the the control culture not exposed to 6-TG, essentially all the cells divide and incorporate thymidine. In the culture exposed to 6-TG, normal cells are killed and therefore do not incorporate the radioactive thymidine into their DNA. The cells from both cultures are then fixed on microscope slides, covered with an X-ray film or emulsion, and then developed using standard radiographic techniques. Cells that have incorporated radioactive thymidine emit radiation, causing a black, exposed spot on the developed film. The developed films are then viewed under a microscope. This technique is quite sensitive; radioactive cells can be counted at a rate of 2 per 10,000 to 1 per 10 million cells. The procedure can also be automated to make it relatively efficient for identifying rare mutants. Dividing the fraction of radioactive cells in the 6-TG-exposed culture by the fraction in the culture not exposed to the drug produces an estimate of the frequency of 6-TG-resistant mutants.

The second method, called the clonal assay, selects for cells that are resistant to 6-TG. White blood cells are dispensed in small test tubes or wells in a small plastic device resembling a tiny muffin tin, each test tube or well containing 6-TG in the growth medium. Only the mutant 6-TG-resistant cells grow and divide, forming visible colonies within 10 to 14 days. These colonies can then be analyzed further by isolating their DNA and analyzing it for mutations. The clonal assay allows characterization of mutations found in these cells, whereas the autoradiographic method permits only determination of their frequency.

Mutational Spectra From Human T-Lymphocyte HPRT Mutants

Cariello (16) described a technique for analysis of the *hprt* gene that may correlate mutational patterns with particular mutagenic exposures. In this method, all of the mutant *hprt*$^-$ T-lymphocytes in a small blood sample are isolated and grown in culture for 2 weeks to produce enough DNA for analysis. DNA of these clones of different *hprt*$^-$ mutants is then examined by first cutting it into fragments with restriction enzymes, adding a group of different DNA probes and separating DNA bound to the probes by gradient gel electrophoresis. DNA with different mutations will bind to different probes. The pattern of mutations in the original blood sample appears as the pattern of DNA probe positions on a gel. Different mutagens are known to produce clearly different patterns of mutation in human as well as bacterial cells, so the aim of this technique is to identify the probable causes of mutations in human blood cell samples by using these patterns, or "mutational spectra" (139). This approach includes studies of the pattern of spontaneous mutation in human T-lymphocyte samples as well as studies of the mutational spectra produced by potentially mutagenic chemicals to which human beings may be exposed. The goal of this approach is to "fingerprint" for the presence of a particular mutagen in an individual's life history.

The *hprt* Gene

One of the advantages of studying the *hprt* gene is that much has been learned about it from studies of a particular genetic disease. Heritable mutations in the *hprt* gene located on the X chromosome lead to Lesch-Nyhan syndrome, a severely debilitating disorder characterized by mental retardation and self-mutilation. The *hprt* gene has been isolated from blood samples taken from patients with Lesch-Nyhan syndrome, and the mutations in the *hprt* gene have been studied. Comparisons have been made between the types of mutations present in cells of Lesch-Nyhan patients and those present in rare, mutant *hprt*$^-$ T-lymphocytes selected from blood samples of people without the disease. DNA from 5 of 28 Lesch-Nyhan patients were found to have major *hprt* gene alterations—deletions of DNA, amplification of DNA, and detectable changes in DNA sequence (172). Studies of mutant T-lymphocytes selected from normal blood samples have shown similar gene alterations in several cases (4,143). These results show that the *hprt*$^-$ T-lymphocyte assay detects spontaneous mutations in the *hprt* gene in somatic cells that are qualitatively similar to clinically apparent heritable mutations in humans, and also demonstrates that the clonal assay in T-lymphocytes detects true mutations, not phenotypic changes that mimic mutations.

Results of Studies of *hprt*$^-$ T-lymphocytes

A number of studies have measured frequencies of T-lymphocytes resistant to 6-TG in normal individuals. Average mutant frequencies for the groups of people tested fall within a fifteenfold range, between about 1 and 15 mutations per million cells (table 7). However, mutant frequencies for different individuals vary widely. For instance, Albertini reports a range of frequencies from 0.4 to 42 mutants per million T-lymphocytes from 23 individuals, representing a hundredfold range in somatic mutant cell frequencies among a small number of people (2). This wide range in

Table 7.—Frequencies of hprt⁻ T-Lymphocytes in Human Beings

Average mutant frequency	Range	Number	Reference
1.1×10^{-6}	NR[a]	12	(70)
2.2×10^{-6}	NR	NR[a]	(1)
2.9×10^{-6}	NR	11	(117)
3.0×10^{-6}	NR	14	(79)
3.8×10^{-6}	—	1	(2)
4.6×10^{-6}	$3.9-6.0\times10^{-6}$	2	(137)
6.1×10^{-6}	NR	45	(26)
1.2×10^{-5}	—	1	(3)
1.4×10^{-5}	$0.4-42\times10^{-6}$	23	(2)
1.5×10^{-5}	$0.8-2.5\times10^{-5}$	24	(160)

[a]Number of individuals not reported.

SOURCE: Office of Technology Assessment.

mutant frequencies may be due to different ages and other characteristics of the T-lymphocyte donors. Two investigators have shown that there is a linear increase of *hprt*⁻ mutants in T-lymphocytes with increasing age (141,160). Donor characteristics, such as smoking behavior, have been shown to increase somatic mutant frequency; genetic variation in sensitivity to mutagens may also exist among individuals.

The degree to which the observed frequency of mutants reflects the actual frequency of somatic mutations in this assay is not fully understood. Descendants of the original mutant cells increase the observed frequency, so growth of mutants in vivo could falsely lead to a conclusion of higher mutation rates than is actually the case, unless the data are corrected for this problem. In addition, *hprt*⁻ cells may be shorter-lived than *hprt*⁺ cells. Before the technique can be used to estimate possible increases in somatic mutation rates from environmental exposures, the wide range of background rates of these mutants has to be investigated, including discrimination between true genetic events and any possible phenocopies. While imprecision of the assay may contribute to this wide range of mutant frequencies, such a range is not unexpected for genetically and geographically diverse human populations.

DETECTION OF SOMATIC MUTATIONS IN RED BLOOD CELLS

Unlike white blood cells, mature red blood cells have no nuclei and contain no DNA. They cannot be grown in the laboratory since they do not have the capacity to replicate. Red blood cells can, however, be used to detect mutations expressed in altered proteins on their surface. Red blood cells are also easily obtained from a blood sample. There are about 1 billion red blood cells per milliliter of whole blood. Branscomb (13) described two methods to detect variant red cells, using high-speed automated microscopy and flow cytometry to analyze and sort hundreds of red blood cells per second to detect rare red cell mutants. These two methods are described below.

Glycophorin Somatic Cell Mutation Assay

The glycophorin assay detects gene-loss mutations expressed by the absence of cellular proteins corresponding to those genes. Such mutations can be detected where two or more variants of a particular protein normally exist, where the genes for the alternative forms are said to be codominant (both variants are expressed when a gene for each is present), and where different variants are expressed in the same cell. A mutation causing the loss of function of the gene for one of the variant forms results in a cell that expresses only the other form of the protein (9,10,50).

One such protein is glycophorin A, a glycoprotein present on the surface of red blood cells. Two variants of glycophorin A normally exist, called "M" and "N," which differ from each other by 2 out of a total of 131 amino acids. The M and N serotypes have no known biological function, and each allele functions independently of the other. The M and N variants of glycophorin A can be labeled independently using monoclonal antibodies carrying different fluorescing molecules. As a result of the labeling, the M variant appears green and the N appears red when viewed under a fluorescent microscope. A flow cytometer, a machine through which cells are passed at very high speed, can sort and count the number of cells that are green, containing only the M variant, and the number that are red, containing only the N variant.

Individuals whose red cells display both variants of glycophorin are heterozygous for the M and N alleles (i.e., they inherited an M gene from one parent and an N gene from the other). As used currently, this assay detects gene loss mutations at glycophorin genes in individuals who are heterozygous for the M and N variants. Red blood cells from heterozygotes show both red and green fluorescence; if a mutation inactivates one of the glycophorin A alleles, only the other variant would be present on the cell surface, hence only one color would appear. Single-color cells are recognized and distinguished from the majority of double-colored cells, and are counted as somatic mutant cells.

The glycophorin A assay is currently being used at the Lawrence Livermore Laboratory to study mutations in two populations: cancer patients' blood cells examined before and after treatment with known doses of chemotherapeutic drugs, and Japanese atomic bomb survivors whose dose of radiation can be estimated (69).

Hemoglobin Somatic Cell Assay

The frequency of particular hemoglobin variants that exist at low frequencies in red blood cells of normal individuals can be measured. Cells containing these variants presumably arise in bone marrow stem cells, the precursors of red blood cells. Methods have been developed using monoclonal antibodies, prepared for each type of variant, to label particular mutant hemoglobin molecules. This method detects small mutations, such as single nucleotide changes resulting in amino acid substitutions, frame shift mutations, etc., in the gene for beta-globin, a constituent of hemoglobin.

Investigators at the University of Washington (132,133) and at Lawrence Livermore National Laboratory (50) have developed an approach to detect hemoglobin mutations in red blood cells. In this technique, an antibody to the mutant hemoglobin is bound to a chemical called a fluorochrome, which is then mixed with red cells in suspension. The fluorochrome emits fluorescent light of a specific color when light of a particular wavelength is shone on it. The cells are screened under a microscope that allows the investigator to direct light of a particular wavelength on the sample. If any of the cells contain the mutant hemoglobin bound to fluorescing antibody, they can be viewed and counted.

Stamatoyannopoulos and his colleagues have used this technique on blood samples from 15 individuals using antibodies specific for Hemoglobin S (the mutant form of hemoglobin present in sickle cell anemia) and Hemoglobin C. An average of 50 million cells per individual was screened, and an average mutant frequency of 1.1 in 10 million cells per subject was obtained, with a range of 4 in 100 million to 3 in 10 million. These investigators also screened the red blood cells of 10 individuals who had been exposed to known mutagens—X-rays or mutagenic drugs used to treat cancer. The frequencies of mutant hemoglobins obtained from these samples were within the range found for normal individuals.

This technique is labor-intensive because it relies on visual screening of the cells: it takes one technician nearly 1 month to screen 100 million red cells. The investigators have since collaborated with Dr. Mendelsohn and his associates at the Lawrence Livermore Laboratory to use their automated fluorescence-activated cell sorter to screen the cells. Although some technical difficulties were encountered with this instrument, results obtained on red cells from six normal individuals (on screening 500 million cells per individual) yielded an average mutant frequency of 1 in 10 million, about the same as the average frequency determined from visually counted cells.

With an array of different mutant hemoglobins to use for production of antibodies, a variety of mutations can be detected in hemoglobins found in rare cells present in normal individuals: single nucleotide substitutions, deletions, frame-shift mutations, or virtually any mutation that causes the production of mutant hemoglobins.

There are currently two technical problems with this technique. First, a significant number of false positives are generated. Better methods of fixation and antibody staining of the hemoglobins are necessary, along with verification of each variant found. Second, the method is not yet fully automated, but research in several laboratories is progressing toward full automation. Significant progress has already been made; Mendelson (68) reported that with present capabilities using flow sorting, 88 antibody-labeled red cells can be identified in a mixture that contains a known quantity of 100 such labeled red cells and 1 billion unlabeled red cells, a relatively good retrieval rate.

CONCLUSIONS

The methods to detect and quantify somatic mutations suggest some additional approaches for studying the nature and mechanisms of mutation in human beings. Somatic tests offer the possibility of drawing associations between exposures to specific mutagens and particular genetic events in the cell, and they may help to identify individuals and populations at high risk for mutations. They may lead to understanding the relationships between the frequency and nature of somatic and germinal mutations, and if those relationships can be identified, somatic tests may contribute significantly to understanding why human heritable mutations occur and may help predict their future occurrence. At present, however, it is not possible to directly extrapolate from mutations in somatic cells to risks of heritable mutations in offspring.

Animal studies can be used to measure quantitatively the correlations between exposure to mutagens and rates of somatic mutation, as well as to investigate possible associations between somatic and heritable mutations. Coordinated tests of somatic and germinal mutations in experimental animals may help guide such research in humans. Such data, combined with data on human somatic mutation rates, may suggest specific risks of heritable mutations in human populations.

In this regard, the tests for somatic mutation described in this report appear promising. None of these methods is currently ready for application to human population studies, however, even though many of the technical details have been worked out. Since the assays do not detect DNA mutations directly, it is particularly important to verify the results from these methods by isolating variant cells and confirming that the altered proteins do indeed reflect primary changes in the genes coding for these proteins. It is also necessary, although more difficult, to determine how often these methods miss the mutations they are designed to detect.

In general, the most useful somatic cell mutation assay may be one that detects a spectrum of mutations (single amino acid substitutions to large deletions and translocations) and that can detect mutations in different cell types. Different tissues may demonstrate different frequencies of somatic mutants, and equally as important, some types of mutant cells may have longer lifespans in the body than others. Consideration of such characteristics could be useful in choosing the most appropriate assay for particular circumstances. Short-lived mutants could be useful in studying the effects of short-term exposures to potential mutagens, while longer lived mutants could be better indicators of mutagenic effects of long-term, chronic exposure to environmental mutagens.

The genetic analysis of white blood cell DNA offers the possibility of identifying genetic "fingerprints," or particular patterns of mutations induced by specific mutagenic agents. Such patterns are an essential first step toward understanding the molecular mechanisms of genetic change in human DNA.

Chapter 6

Laboratory Determination of Heritable Mutation Rates

Contents

TABLE

FIGURE

Chapter 6

Laboratory Determination of Heritable Mutation Rates

INTRODUCTION

The genetic information of all organisms is encoded in DNA (with the exception of some viruses that depend on RNA), and mutations have been found in all species that have been examined. The biological similarity of DNA structure and function in all species has allowed scientists to study mutation rates and mechanisms of mutagenesis in several species for over half a century. In addition to these efforts in basic research, in recent years the biotechnology industry has provided impetus, including financial incentives, for sophisticated inquiry into mutations and mutagenesis in bacteria and yeast. The industry's interest is in using mutagenesis to benefit from some of the particularly useful characteristics of certain microorganisms, such as the ability of some bacteria to digest oil, and the fermentation capabilities of various yeasts.

Another reason for studying mutations in lower organisms is to identify agents that cause mutations in their DNA, agents that might act similarly on human DNA. For instance, concern about human risks from the genetic effects of ionizing radiation led to the funding of extensive mouse studies, which have provided quantitative data on factors that influence the yield of various types of heritable mutations. In the case of chemicals, however, concern about cancer, not about heritable mutations, has been the primary motivation for developing short-term laboratory test systems to detect mutations. In the last decade, as more and more evidence incriminates somatic mutation as an early event in the development of cancer, the properties of mutagenicity and carcinogenicity have been conjoined, making identification of mutagens an important part of efforts to identify cancer-causing agents. Many of the tests for mutagens measure a chemical's capacity to alter the DNA of bacteria or yeast (145). While such a short-term test in a lower organism may distinguish between agents that can and cannot change DNA, it provides no information for making quantitative estimates of mutagenic potency in humans nor does it show that the chemical can necessarily reach animal or human gonads in an active form to cause heritable mutations (108). Whole animal tests specifically designed to detect germ-cell or heritable mutations provide some information of that type.

RODENT TESTS FOR HERITABLE MUTATIONS

At least 15 tests and variations on them are available for the detection and measurement of heritable mutations induced in mammalian germ cells. Lower organisms, notably Drosophila, are the subjects of similar tests, but most of the information about mutation rates comes from mice. The three most widely used rodent tests are described below (58,108).

The Dominant Lethal Test

The dominant lethal test detects mutations that cause the death of rodent embryos. The usual procedure in this test is to treat male mice with a suspected mutagen and then mate them with untreated females. If the treatment causes a mutation in a male germ cell that is lethal at some stage in the development of the embryo (after fertilization), the lethal events are detectable by dissecting female uteri a few days after implantation of embryos (implantation occurs about nine days after fertilization). If there are more dead embryos than expected (compared with matings of unexposed mice) and/or more implantation sites than expected in the uterus where there is no longer an embryo, the effect is assumed to be the result of a mutation in the male germ cell. Chromosomal studies in mice have shown that almost all "dominant lethal" events are associated with major chromosome abnormalities.

Human analogs of the postimplantation fetal losses counted in dominant lethal tests are spontaneous abortions (miscarriages). Fifty to 60 percent of spontaneous abortions are associated with numerical and structural chromosome abnormalities, the same basic types of abnormalities detected by dominant lethal tests (12). These similarities between mice and men, and the general pragmatic acceptance of higher animals as predictors for human risk, underlie the idea that agents that are positive in a mouse dominant lethal test may also cause spontaneous abortions or chromosome abnormalities in liveborn infants.

The Heritable Translocation Test

The heritable translocation test detects breakage and rejoining of fragments from different chromosomes. When a translocation involves no loss of genetic information (a "balanced translocation"), often it has no adverse effect on the functioning of the carrier individual. The unusual chromosomes produced by the breaking and rejoining are present in all cells, and they can pair with each other and divide and function like normal chromosomes. However, such translocations cause reproductive problems because the germ cells may have incomplete sets of genetic material and may lead to unbalanced chromosome abnormalities in offspring, often resulting in death during embryonic or fetal development. Heritable translocations are thus detectable in mice because they result in reduced litter size in the second generation after exposure to the suspect agent, or in sterility in the first generation. In addition, cytogenetic studies can detect abnormal chromosomes and can be used to study the association between reduced fertility and chromosome abnormalities.

Specific Locus Tests

Since 1927, when Herman Müller first demonstrated that radiation causes mutations in fruit flies (81), there has been concern about possible genetic effects of radiation in human beings. Following World War II, in anticipation of widespread use of nuclear power, the Atomic Energy Commission (AEC) contracted for research on the mutagenic effects of radiation in mice to provide an experimental underpinning for estimating effects in human beings. Specific locus tests that detect mutations in a few specific genes in mice were developed for that purpose. This work and use of specific locus tests to measure chemically-induced mutations have provided the animal data most useful for estimating human risk.

Through research supported by the AEC, William L. Russell (109) bred a strain of "tester" mice that differs from wild-type mice in 7 morphological features—coat colors, patterns of pigmentation, eye color, and ear shape—that are recessive traits. Each of the features is thought to be controlled by a different genetic locus. Tester mice are homozygous for the recessive alleles of these genes, while wild-type mice are homozygous for the dominant alleles. When the tester strain is mated to a strain of wild-type mice, the offspring should all be heterozygous at the seven test loci, and therefore they should appear wild-type because the the wild-type alleles are associated with the dominant phenotype. However, if a germinal mutation occurs in a gene for one of the seven loci in the wild-type strain, causing the dominant allele not to be expressed, the offspring from that mutation will have a phenotype characteristic of the tester strain.

In a specific locus test, the wild-type mouse is exposed to radiation or to a known or suspected chemical mutagen and then mated to the tester strain. In most cases, only male wild-type mice are exposed, and the tester strain mice are female. The reason for this is largely a practical one: one exposed male can sire a large number of progeny when mated to many females, but an exposed female can have only one litter at a time. In addition, the information so far available suggests that male germ cells are more sensitive to mutagens than are female germ cells. After mating exposed wild-type and tester mice, the offspring are examined for the easily seen morphological features related to the seven loci. Since the morphologic variants are readily identified, large numbers of offspring can be examined quickly by an investigator. The frequency of mutant offspring of exposed parents is compared with offspring of unexposed parents. The importance of the ease of recognizing mutants is underscored by the low frequency at which mutations occur spontaneously —as few as 50 mutations per 1 million offspring of unexposed mice. (Because mutations can be detected at any of the seven loci, examination of one mouse provides information about one observation at each of the seven loci.)

Estimates for the rate of spontaneous mutations in mice based on specific locus test results are summarized in table 8. These data are based on examining almost 1 million progeny of exposed mice, or about 7 million loci. Though the reliability of the estimate is limited by the small number of mutations seen, the calculated rate of between two and eight mutations per 1 million genes is not inconsistent with the human rate estimated by one-dimensional electrophoresis of blood proteins (see ch. 3).

Although the "average" rate of mutations is a useful touchstone, it is used in the full knowledge that the average includes very different rates at the seven loci. For instance, there is a 35-fold difference in radiation-induced mutation rates between the most and least sensitive of these particular seven loci. The choice of a different set of loci for examination might yield quite different results.

Other Studies in Animals

There is a family of animal experiments called dominant mutation tests, which use the appearance of dominant features, such as skeletal anomalies or cataracts, to detect mutations (108). Many of the endpoints used resemble known human genetic disorders, and the results of these tests have been used to estimate human genetic risk.

In addition to the types of studies described above, some of the techniques used to study human beings have been applied to experimental animals. Electrophoresis of blood proteins and quantitative enzyme assays in human beings, described in chapter 3, have been used to study corresponding proteins in mice. The electrophoretic approach has been used in studies of mutagenized mice at the National Insitute of Environmental Health Sciences (51,52,53,63) and also at the Institute for Genetics, Neuherberg, Germany (19,19a). A small effort in electrophoretic analysis has also been conducted at the Oak Ridge National Laboratory (107). Similar studies have been reported in Drosophila (80,103,140,161).

Mutagenized mice have been tested for enzyme deficiency variants using quantitative enzyme assay techniques similar to those used to study human populations (19a,73). Enzyme assays have also been used in mouse and Drosophila experiments that involve selective mating, similar to the specific locus test strategy. By selecting parental strains that have electrophoretically distinct alleles at a locus, enzyme deficiency variants can be detected in offspring by the absence of one activity staining band that, in the absence of a

Table 8.—Spontaneous Mutation Rates in Mice[a]

Sex	Number of mutants	Number of progeny[a]	Mutations/locus[b]	Reference
Female	3 or 8[c]	204,639	2.1 or 5.6 $\times 10^{-6}$	(110)
Male	39	727,319	7.6 $\times 10^{-6}$	(113)

[a]Data about spontaneous mutation rates were collected from animals used as controls in experiments to determine the effects of radiation or chemical agents on males and females separately. There are more progeny from males because more experiments have been done in males.
[b]A rate of 1 $\times 10^{-6}$ means that, on average, a mutation occurs once in a million loci.
[c]The choice of three or eight mutants exemplifies a problem in counting mutations. Three litters from female mice had mutant offspring. Two litters had one mutant each; the third had six mutants, all of which may have resulted from a single mutation. The result is three probable mutations and eight mutant offspring.

SOURCE: Office of Technology Assessment.

mutation, would be inherited from one of the parents (51,52,103,161).

Data from animal experiments using electrophoretic and quantitative enzyme assays are limited, but at this stage a general, cautious conclusion can be drawn. The background mutation rate in the mouse derived from these experimental data is in the same range as the estimate of the spontaneous rate derived from studies in human beings (77), recognizing, however, that these estimates are based on small numbers and may be heavily influenced by chance fluctuations.

FACTORS THAT INFLUENCE INDUCED MUTATION RATES IN MICE

When radiation or a chemical is tested for mutagenicity, the experiment does not necessarily produce definitive answers to the questions: "Is this agent mutagenic?" And "if so, how mutagenic is it?" The answers differ depending on several factors, two important ones being whether the animal is male or female and the stage of development of the exposed germ cell. There is ample evidence from radiation studies that male and female germ cells differ in their sensitivity to mutagens. In males, germ cells appear to be more likely to sustain a mutation the further along the developmental path they are. Overall, female germ cells appear to be more resistant to mutagens than are male germ cells.

Effects in Males

The male testis, from puberty on, contains a mixture of germ cells at different developmental stages, including spermatogonial stem cells, differentiating spermatogonia, and intermediate stage cells, the spermatocytes and spermatids, which are maturing into spermatozoa. The mechanisms of sperm production are qualitatively very similar in mice and men, differing largely in timing: in mice, sperm production from spermatogonia to spermatozoa takes 35 days; in men, it takes 74 days (60). When a male is irradiated, germ cells at all stages of development are exposed. By varying the times of mating, the effects of radiation on different stages can be examined. For instance, if male mice are irradiated and mated immediately, any mutants that appear in the progeny must result from mutations that were caused in fully mature sperm. In a mating two weeks after exposure, eggs are fertilized by spermatozoa that were spermatids at the time of irradiation. If seven or more weeks are allowed to pass, sperm in the ejaculate would have been irradiated at the spermatogonial stem-cell stage. Experimental evidences shows that cells at each stage have varying sensitivities to induction of mutations by radiation, the later, postspermatogonial stages being most sensitive.

While spermatogonial stem cells are more resistant to the effects of mutagens than are germ cells in later stages of development, mutations in stem cells are of greater concern. Stem cells persist throughout the lives of males, constantly giving rise to new generations of sperm cells. A mutation in stem cell DNA results in every cell that buds off from the stem cell bearing a mutation. The lifespan of later germ-cell stages constitutes only a small fraction of the total reproductive lifespan (especially in humans). Thus, in the case of *acute* exposure to these later germ-cell stages, only conceptions occurring during a short postexposure interval are at risk. For these reasons, the effects of radiation and some chemicals have been especially closely studied in spermatogonial stem cells in animals.

The most common way for people to be exposed to ionizing radiation is through acute exposure to medical X-rays, or long-term, low-level exposure to natural background radiation. Certain groups are exposed occupationally (e.g., uranium miners, nuclear powerplant workers, some medical personnel); and some may have special low-level environmental exposures from living near nuclear installations, nuclear waste sites, uranium mine tailings, or areas of high natural radon concentrations.

In mice, exposure of males to X-rays or other forms of ionizing radiation at low dose rates causes a measurable increase in the number of mutations, while such exposures do not measurably increase mutation rates in females. Russell and Kelly (113) counted the numbers of mutants caused by varying doses of radiation in male mice. They find no evidence for a dose so low that it is not mutagenic, but the number of mutations decreases with lower and lower doses (see fig. 13). The data in the figure also show the differing results of exposure at high and low dose *rates*. For example, the number of mutations caused by a total dose of 300 roentgens (R),[1] administered at a dose rate of between 72 and 90 R per minute is about 80 X 10^{-6} per locus; the same *total* dose administered at dose rates of 0.08 R per minute or lower causes fewer mutations, about 30 X 10^{-6} per locus. These data may be interpreted by saying that low dose rates allow time for repair of some of the damage caused by the radiation. (For further discussion of dose and dose-rate relationships, see ch. 7.)

Figure 13.—Spermatogonial Mutation Rates in Mice After Low-Dose Rate and High-Dose Rate Radiation

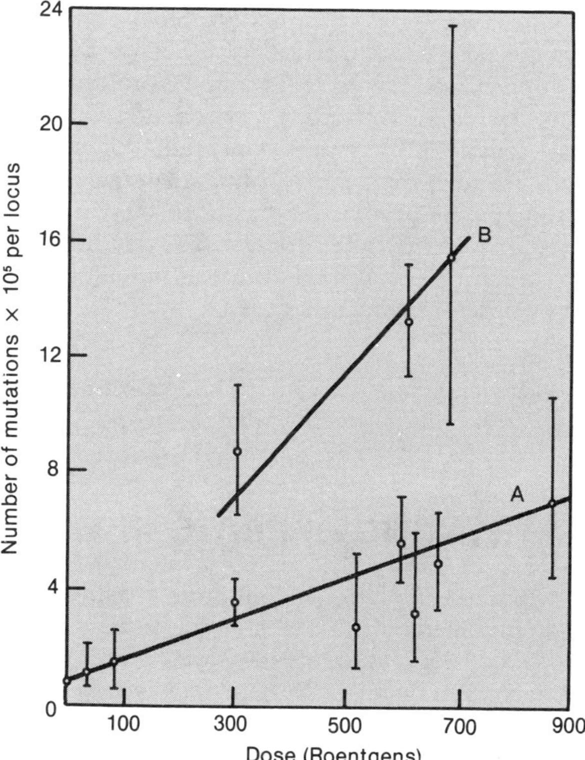

Straight lines fitted to experimental results from a number of specific-locus tests involving single doses of radiation. **Line A:** Radiation delivered at low dose rates (0.8 Roentgens/minute and lower). **Line B:** Radiation delivered at high dose rates (72 to 90 Roentgens/minute). The figure shows that a dose of radiation delivered at a faster rate causes more mutations per locus than the same total dose of radiation delivered at a slower rate. (90% confidence limits are shown.)

SOURCE: W. L. Russell and E. M. Kelly, "Mutation Frequencies in Male Mice and the Estimation of Genetic Hazards of Radiation in Men," *Proceedings of the National Academy of Sciences* (U.S.) 79:542-544, 1982.

Effects in Females

Female mammals differ from males in the biology of germinal stem cells. Whereas male spermatogonial stem cells divide constantly during adulthood, females are born with their total complement of egg precursor cells. Shortly after birth, all cells destined to become eggs (ova) are present in the ovaries as "arrested oocytes." Exposure of arrested oocytes to radiation has caused no increase in mutation rates in mice (110). During the six weeks before ovulation an arrested oocyte matures, undergoing meiosis and, in a mature state, travels from the ovary to the uterus. Irradiation during this maturation period does increase mutation rates in female germ cells.

Two different explanations can be advanced for the failure to recover mutations from irradiated arrested oocytes. First, it might be that arrested oocyte DNA is somehow protected from the mutagenic effects of radiation, perhaps by having very efficient repair mechanisms that correct radiation-induced mutations. On the other hand, in mice, arrested oocytes are very sensitive to the *killing* effects of radiation (27). These lethal effects could also account for the absence of mutagenic effects in female mice. It is argued that since radiation kills oocytes very efficiently, oocytes would either die before they were mutated or they would die with mutations that would never be expressed.

The extreme sensitivity to the killing effects of radiation in mouse arrested oocytes is not common to all animals. In the squirrel monkey, high sensitivity is limited to oocytes irradiated during fetal life. Arrested oocytes in other monkey species are no more sensitive to lethal effects of radiation than are other cells. Similarly, human oocytes, at least those that are irradiated after

[1] A measure of radiation. In human terms, a chest X-ray examination produces a gonadal radiation dose of less than 0.0005 R (148).

birth, appear resistant to the lethal effects of radiation.

The differing sensitivities of oocytes of different species to lethal effects of ionizing radiation complicate efforts to extrapolate from mutation rates in female mice to the expected rates in human females (28). In particular, radiation may result in more mutations in human females than would be predicted from female mice because human arrested oocytes are less likely to be killed by radiation. That might allow a human oocyte to survive a mutagenic dose of radiation and to be fertilized.

Using experimental data from female mice, Russell (1977) made several calculations relating effects of radiation in mice to potential effects in human females. In each calculation, he made the "worst case" assumption that all stages of female germ cells were as sensitive to the mutagenic effects of radiation as are the most sensitive stages—maturing and mature oocytes—in mice. Because there are different methods of handling the data, he made four different calculations. Three produced estimates indicating that radiation would not increase the mutation rate in human arrested oocytes. The fourth estimated that human arrested oocytes would be somewhat less than half as sensitive (from 17 to 44 percent as sensitive) to radiation as are human spermatogonia.

ESTIMATING EFFECTS IN HUMAN BEINGS FROM ANIMAL DATA

Most extrapolations from animal data to human mutagenic risks are based on studies in male mice because: 1) there are more data for male mice than for female mice, 2) there appear to be fewer biological differences in germ cell development between males of the two species than there are between females (see above), and 3) male germ cells appear to be more sensitive to radiation than female germ cells and may provide a more sensitive indicator of genetic risk.

The dose-response curves in figure 13 are derived from experiments on male mice and they provide information for estimating human risks. The absence of a detectable threshold for genetic effects in male mice suggests that humans are at some increased risk for mutations at any level of radiation exposure. This finding increases the importance of making quantitative estimates of the effects of radiation; since risk does not go to zero except at zero dose, what is it at levels of human exposure? The "doubling dose," which is the dose of radiation that induces an additional number of mutations equal to the number that occur spontaneously (resulting from all mutagenic influences, known and unknown), can be derived from the information in figure 13. The doubling dose differs depending on the radiation dose *rate*. The doubling dose in male mice for high dose rates is about 40 R; for low dose rates, about 110 R. This is of practical importance for human beings, since some exposures are at high dose rates—e.g. the populations of Nagasaki and Hiroshima and people receiving radiation therapy for some types of cancer—but most population exposure to radiation is delivered at a low dose rate—e.g., people who live near radioactive mines. About three times as many mutations are caused by radiation delivered at a high dose rate as opposed to an equal total radiation dose delivered at a low dose rate (114).

In chapter 3, the effects on mutation rates of acute irradiation of the populations at Nagasaki and Hiroshima are discussed. The exact doubling dose for those populations is still being debated because of uncertainties about the total dose and because of insensitivity of the methods used to detect mutations. Using the best information available about those populations, however, humans appear to be less sensitive than mice to high dose rate exposures.

Current radiation exposures of United States citizens average about 200 millirem (mrem) per year. On a population basis, this exposure is generally divided about equally between high dose rate medical exposures and low dose rate exposures to natural sources of radiation. These exposures do not approach the doubling dose for mutations. If the average rate of human mutations is now on the order of one mutation per 1 million genes per generation, we can estimate the effect of a hypothetical increase in human radiation exposure to twice the current level. If exposures increased to 400 mrem and if that exposure were as mutagenic as the high dose rate delivered to mice, it would increase the average human rate to 1.005 mutations per 1 million genes per generation. Such an increase would be undetectable by any current method.

Animal studies have proven useful for learning about the various relationships of dose, particularly for radiation, and germ-cell and heritable mutation rates. They are limited, however, in providing information directly applicable to human beings. First and most important, human beings differ from animals in ways that we know about and in more ways that are not understood. Second, the endpoints measured in animal tests do not generally correspond directly to endpoints measurable in human beings under nonexperimental conditions. Third, the available animal tests, like the currently available methods for studying human beings, detect mutations arising in specific, selected, loci which may not be representative of other loci. Evidence from Russell's specific locus test, cited above, shows a 35-fold difference between the least sensitive and most sensitive of the loci tested. A fourth consideration, again similar to the limitations of studies in humans, is that each test does not detect more than a fraction of the mutations that can occur, and the exact nature of the mutations that are detectable cannot be characterized with these methods.

When the new technologies discussed in this report are further developed, they presumably will be applied to experimental animals as well as to human beings. Studies that examine corresponding genes with the same techniques in humans and animals may provide some of the interspecies comparisons that now are lacking.

Chapter 7

Extrapolation

Contents

TABLE

FIGURES

Chapter 7
Extrapolation

INTRODUCTION

Currently, predictions of possible risk of heritable mutations in human beings are based on inferences, or "extrapolating" results, of mutagenicity tests in other organisms or in laboratory cell cultures. One of the key problems in genetic extrapolation is that, while there is no shortage of mutagenicity tests using a variety of organisms and cell types, researchers have just begun the painstaking work of drawing out relationships among results of different tests, and of eventually validating models for extrapolating to heritable effects in human beings (see fig. 14). This chapter reviews the basic constructs that have been devised for framing genetic extrapolations and then presents some efforts that have been made to carry out specific extrapolations.

The Aims of Extrapolation

For many years, the emphasis in mutagenicity testing has been on developing individual tests and learning about their properties. Test development has not been targeted exclusively toward developing predictive models for heritable mutations in human beings and in fact the quest for tests that could be used to predict carcinogenicity through somatic mutation has predominated. This being the case, the test systems vary tremendously and include, for example, tests in bacteria, insects, animal somatic and germ cells in culture, whole mammals, and human somatic cells. Efforts to relate results from one test system to another or from various tests to human beings, even on a qualitative level, have been relatively recent and have not progressed very far to date. Bridges (15) summarized this situation:

> Despite the extremely large number of screening tests that have been performed, remarkably little rational thought has been devoted to the use that should be made of the results of such tests. Mutagenicity tests in lower organisms[1] are essentially screening tests warning of a possible human hazard. They do not give any quantitative indication of the level of risk to man, not so much because of differences in the organization of DNA in man and lower organisms, but because of the complex metabolic capabilities of man which may greatly enhance or diminish the mutagenic effectiveness of an agent.

[1]"Lower organisms" refers to bacteria, yeast, Drosophila, etc., and not to whole mammals.

In recent years, scientists who use different experimental approaches to study mutations have begun to discuss methods to correlate results from various test systems. The first consideration is whether results from one sytem are *qualitatively* predictive of results in another, i.e., are appropriate endpoints being considered and do the results agree in whether they are positive or negative. This is called "biologic extrapolation." The second type of extrapolation is *quantitative*, which deals with the relationship between the quantitative response in a test to a quantitative estimate of the likely effect in human beings.

Examples of questions posed in extrapolation are the following: If experiments demonstrate that a single exposure to a high dose of a chemical induces heritable mutations in mice, what would be the result of a single exposure to a lower realistic dose or of a long-term exposure to a lower dose of that chemical in humans? How do results of mutagenesis experiments on rodent somatic cells in a test tube cell culture (e.g., Chinese hamster ovary cells "in vitro") relate to the issue of mutation rates in human somatic cells, or mutation rates in human germ cells? What are appropriate dose conversions between various experimental systems and human beings? If a worker in a chemical plant has a tenfold increase in mutation frequencies in his or her somatic cells, what are the long-term health implications? Is that worker at increased risk of having a child with a new mutation? These types of questions illustrate the complex issues that arise in using results from one system to make predictions of risks to human health.

Figure 14.—Biological Test Systems for Studying Mutations

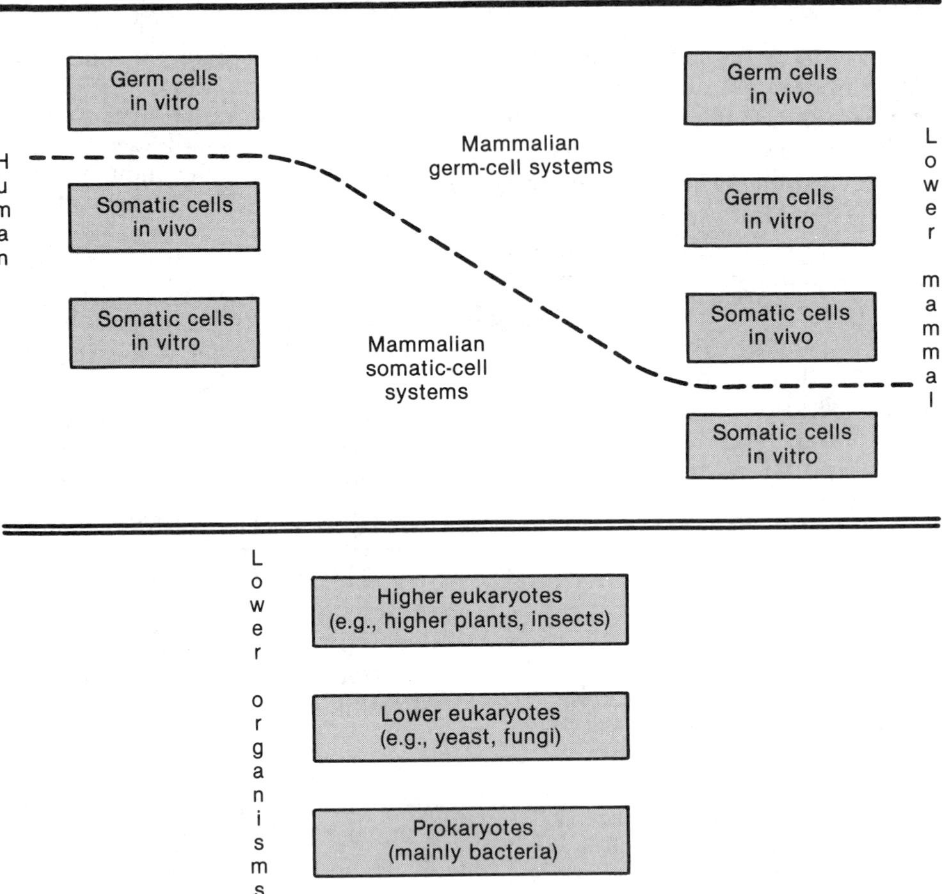

Test systems for studying mutations, arrayed according to biological similarity to the endpoint of interest—heritable mutations in human beings. Some test systems can be used to measure a single endpoint; more than one type of endpoint (e.g., both chromosomal and gene mutations) can be measured in other systems. Extrapolation from any system to human germ cells in vivo involves many, mostly untested, assumptions.

SOURCE: Office of Technology Assessment.

In translating results from one system to another (using similar genetic endpoints), a number of separate extrapolations may have to be made: 1) from species to species; 2) from experimental doses to actual environmental doses; 3) from one cell type to another; 4) from in vitro to in vivo physiological conditions; and 5) the biggest and most uncertain leap, from estimates of mutation frequencies to estimates of genetic disease in humans. The kind of information that would give the biggest boost to the ability to predict effects in humans with information from other test systems is knowing exactly what kinds of mutations (e.g., point mutations, chromosomal rearrangements, etc.) each of the tests detects. The new technologies discussed in this report may provide this type of information.

EXTRAPOLATION MODELS

Several researchers have begun developing models for extrapolating from one test system to another (14,15,46,49,61,64,106,129,130,138). One of the key features common to all the extrapolation models developed is that a result from a single test system would not be used alone to predict a result in another test system. Instead, results from several related test systems are correlated and used together.

Several of the methods described are extensions or rearrangements of the first extrapolation method developed by Sobels (1290)—the parallelogram (130):

> The underlying principle is to obtain information on genetic damage that is hard to measure directly, for example mutation in mouse germ cells, by comparison with endpoints that can be determined experimentally, e.g., alkylation per nucleotide in mammalian cells in vitro and in mouse germ cells, and mutation induction in mammalian cells in vitro.

Sobels' parallelogram is illustrated in figure 15. It is relatively easy to determine the mutation frequency in mouse somatic cells (a quantity called "A") upon exposure to a particular chemical mutagen in culture. With certain types of chemicals (alkylating agents), it is also possible to derive a measure of the interaction of the mutagen with the DNA of those cells, which is quantified as "alkylations per nucleotide" (a quantity called "A"). Alkylations per nucleotide can also be measured in mouse germ cells after exposure to the same mutagen (B'). The relationship of these different values is then used to calculate the expected mutation frequency in mouse germ cells (B) on exposure to the same mutagen. A major, untested assumption is that the ratio of A to A′ is proportional to the ratio of B to B′, i.e., A/A′ = B/B′. If that is true, it is then a matter of simple algebra to predict the mutation frequency in mouse germ cells (B) by solving the equation for B, which is the only unknown quantity.

Figure 15.—An Example of a Parallelogram

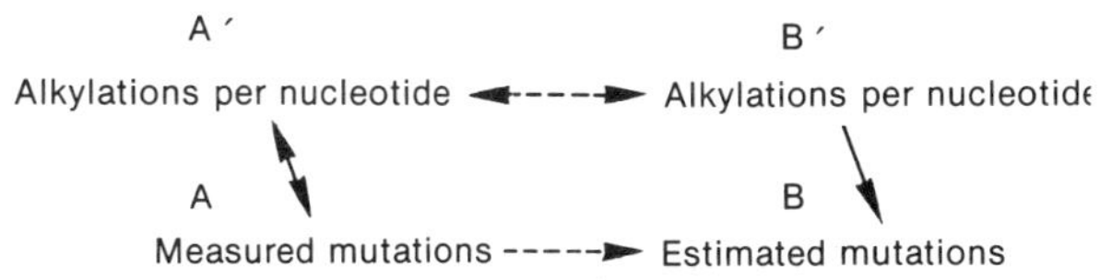

(Note that the same test system is represented in the vertical dimension and the same genetic endpoint in the horizontal dimension.)

SOURCE: F.H. Sobels, "The Parallelogram: An Indirect Approach for the Assessment of Genetic Risks from Chemical Mutagens," pp. 323-327 in K.C. Born et al. (eds.), *Progress in Mutation Research*, vol. 3 (Amsterdam: Elsevier, 1982).

A similar parallelogram can be used to extrapolate results from mouse germ cells to human germ cells. Mutation frequencies are measured in somatic cells of both humans and mice. Germ-cell mutation frequencies are measured in the mouse and compared to somatic-cell mutation frequencies in the mouse. Assuming that the ratio of germ-cell to somatic-cell mutation frequencies is the same in mice and human beings, germ-cell mutation frequencies in human beings can be predicted from the measured human somatic-cell mutation frequencies (64).

Streisinger (138) has proposed a more complex extrapolation method using two sequential parallelograms. In the first parallelogram, a measure of chemical interaction with human germ cell DNA is estimated from measurements of the effects of the chemical in human and animal (e.g., monkey) somatic cells, and a measure of interaction of the same chemical with germ-cell DNA in the same animal. In the second parallelogram, the ratio of a measure of chemical interaction with mouse germ-cell DNA to mouse germ-cell mutation rates (from the specific locus test) is used to predict the human germ-cell mutation rate using the estimated value of chemical interaction with human germ-cell DNA from the first parallelogram. This construct embodies several untested assumptions.

Bridges (1980) also developed a more complex extrapolation model based on the original parallelogram model of Sobels. Based on his approach, Bridges outlined the types of results needed to fill information gaps, ultimately to assess the impact of mutagens at the: 1) molecular; 2) cellular; and 3) whole organism levels in both animals and man. He suggested studies to determine: 1) the presence of an effective dose of mutagen at the molecular level by measuring the concentration of mutagen in the gonads or blood or the extent of reaction with DNA; 2) whether there appears to be a relationship between the presence of the mutagen and a biological response at the cellular level by measuring somatic mutation frequencies or chromosomal changes in lymphocytes; and 3) whether there is an effect at the whole organism level by measuring the frequency of heritable genetic defects, congenital malformations, or fetal loss.

The values Bridges specifies are obtainable in animal systems. To obtain such values for man, Bridges suggests that use be made of certain otherwise normal human populations that are exposed to large doses of mutagens. Examples of such populations are patients treated for diseases, such as cancer, with drugs that are known to be mutagenic, and certain occupational cohorts in which there are known excesses of cancer (1,155). Simultaneous studies using the same mutagens could be carried out in experimental mice to determine the relative sensitivities of mouse and man to these mutagens.

Parallelogram models are attractive for their simplicity and inherent logic. They are appropriate starting points for exploring relationships among test results when sufficient data become available to do so. However, the assumptions embodied in parallelogram models—consistent, predictable relationships among various cell types, translatable among species—are almost entirely untested. The great differences among species make it unlikely that these parallelogram models will survive validation studies intact. While they may continue to be useful research tools for posing logical questions, they may or may not prove practical for predicting risks of heritable mutations in human beings.

ATTEMPTS AT QUALITATIVE EXTRAPOLATION

L. Russell and colleagues (106) compared the results of mutagenicity tests carried out in a variety of systems other than whole mammals with results from specific locus tests (SLTs) and heritable translocation tests (HTTs) in mice (see ch. 6 for descriptions of these two tests). The purpose of the comparison was to find out how well results of the nongerm-cell tests corresponded to the qualitative results (positive or negative) of the two germ-cell tests. The analysis was limited by the relatively small number of chemicals that have been tested in either the SLT or the HLT. About 35 chemicals have been tested in one or both of the germ-cell assays, out of a total of about 2,000 chemicals for which some test results are available from any system.[2]

The comparison tests were grouped into 18 categories and the categories given relative weights according to their biological relationship to one of the germ-cell tests. The categories and their weights are given in table 9. The lowest weighted category includes tests using prokaryotes, such as bacteria, directly treated by the suspected mutagen. Higher scores signify moving toward higher

[2]These test systems are not the focus of this report and are not described here in detail. Descriptions of these tests can be found in (88).

Table 9.—Weighting of Test Results for Presumption of Germ-Line Mutagenicity

	Exposure not within mammalian body	Weight	Exposure within mammalian body	Weight	Germ cells	Weight
Prokaryotes, all endpoints	SAL WP_	1	BFT HMA	2		
Lower eukaryotes, all endpoints	YEA YEP ASP NEU	2				
Higher eukaryotes, chromosome aberrations	PYC	3			DAN DHT	8
Higher eukaryotes, gene mutations	PGM	3			SRL	8
Mammals, genetically nondefined endpoints	SC1 SC2 UDH UDP	4	SC3 SC4	6	SPM	4
					UDT	8
Mammals, chromosome aberrations	CYC	5	MNT	7	DLT CY9 CYO	15
			CYE CYB CY5 CY8	10		
Mammals, gene mutations	CHO V79 L51	5	MST	10	SPF	15

NOTE: In general, the weights increase from top to bottom and from left to right in the table. From top to bottom, the tests progress from lower to higher organisms and from more general endpoints to endpoints of direct relevance to human heritable mutations. From left to right, the categories progress from in vitro tests in both somatic and germ cells, to in vivo germ-cell tests.

Explanation of test symbols:

ASP Aspergillus, all tests
BFT Body fluid tests, all assays
CHO Chinese hamster ovary cells in culture
CYB Mammalian cytogenetics, in vivo, animal bone marrow
CYC Mammalian cytogenetics, in vitro, all cell types
CYE Mammalian cytogenetics, in vivo, animal lymphocytes or leukocytes
CYO Mammalian cytogenetics, in vivo, oocyte or early embryo studies
CY5 Mammalian cytogenetics, in vivo, human bone marrow
CY8 Mammalian cytogenetics, in vivo, human lymphocytes or leukocytes
CY9 Mammalian cytogenetics, all male germ-cell studies
DAN Drosophila aneuploidy studies, all tests
DHT Drosophila heritable translocation test
DLT Dominant-lethal test in rodents
HMA Host-mediated assay studies
L51 Mouse lymphoma cells in culture, gene mutations at TK locus
MNT Micronucleus test, all species
MST Mouse spot test
NEU *Neurospora crassa*, all tests
PGM Plant gene mutations, all tests
PYC Plant chromosome studies, all tests
SAL Salmonella histidine reversion tests
SC1 Sister-chromatid exchange, human cells in vitro
SC2 Sister-chromatid exchange, animal cells in vitro
SC3 Sister-chromatid exchange, animals in vivo
SC4 Sister-chromatid exchange, human cells in vivo
SPF Sperm abnormalities in F_1 males
SPM Sperm abnormalities in treated animals
SRL Drosophila sex-linked recessive lethal test
UDH Unscheduled DNA synthesis, human diploid fibroblasts
UDP Unscheduled DNA synthesis, rat primary hepatocytes
UDT Unscheduled DNA synthesis, testis in vivo
V79 Chinese hamster lung (V79) cells in culture, all gene mutation studies
WP_ *E. coli* reverse mutation studies
YEA *Saccharomyces cerevisiae*, all tests
YEP *Schizosaccharomyces pombe*, all tests

SOURCE: L.B. Russell, C.S. Aaron, F. de Serres, et al., "Evaluation of Mutagenicity Assays for Purposes of Genetic Risk Assessment," *Mutation Research* 134:143-157, 1984.

mammals, toward germ cells, and toward treatment with the chemical in a whole mammal.

A single composite score was calculated for each chemical tested, adding together scores from each category in which there were test results. There is only one score per category regardless of the number of tests. Positive results are scored as positive numbers; negative results as negative numbers, e.g., an in-vitro somatic-cell chromosome aberration test with a positive result yields a score of +5, one with a negative result, a score of −5.

Russell and colleagues found that nearly all chemicals that tested positive in either or both the SLT and HTT had high composite scores from other tests. A number of chemicals with negative SLT and HTT results also had high, positive composite scores, representing "false positives" in the comparison tests.

Similar analyses looked separately at the SLT and HTT and the comparison tests that specifically detect gene mutations or chromosome aberrations, respectively. The results are similar to those matching the results in all comparison tests against both mammalian germ-cell assays: high scores for most chemicals positive in the germ-cell tests, and a number of false positives.

In an additional analysis, the comparison tests are ranked according to how well each predicts the results of the two germ-cell tests. In general, the tests in higher numbered categories in the earlier analyses, i.e., those that are closer biologically to whole mammal germ-cell tests, had better correlations with the SLT and HTT. For the SLT, the best predictors overall were the mouse spot test, unscheduled DNA synthesis in mouse testis, and the micronucleus test. For the HTT, unscheduled DNA synthesis in testis, the dominant-lethal test, and one lower ranked test, sister-chromatid exchange in cultured animal cells, were the strongest predictors.

While it appears that the results of some of the comparison tests correlate relatively well with the mammalian germ-cell tests, in fact, *not one* of these correlations reaches conventional statistical significance, meaning that the tests do not predict reliability better than chance. The lack of significant results is due, in large part, to the small number of comparisons for many tests, and in part because of the process for selecting chemicals for SLT and HTT. From a practical, public policy standpoint, this is an important finding. The lack of statistically significant results does not mean that these comparisons are without value. The study provides a status report on the quality and quantity of existing data.

Since the two whole mammal tests (the SLT and HTT) are relatively expensive and time-consuming, they are usually reserved for testing chemicals highly suspected of causing heritable mutations. The suspicion is based on results of other tests, specifically the comparison tests examined in Russell and colleagues' analysis. It is hardly surprising, therefore, that the comparison test results are largely positive for chemicals eventually tested in the mammalian germ-cell assays. Russell and colleagues took the preponderance of positive results into account in their analyses.

Many chemicals have been tested in mutagenicity assays because, for reasons of chemical structure or other properties, there is a high likelihood that they will be mutagenic. While these chemicals have proved useful as laboratory tools, they are not necessarily useful for drawing conclusions about what people are actually exposed to. W. Russell (111), using the same database used by L. Russell and colleagues (106), looked exclusively at the 11 "environmental chemicals" (those found in the home or workplace) that have been tested in the SLT and examined the results. All 11 are positive in the Drosophila sex-linked lethal test, the 10 that have been tested are positive in mammalian somatic-cell tests, and there are a variety of positive results in other test systems. None of the 11, each of which was tested at very high doses, is positive in the SLT, suggesting no increase in mutations in spermatogonia (the pre-meiotic male germ-cell stage) although several have positive results in tests of later sperm developmental stages.

What conclusions can be drawn about the validity of qualitative extrapolation from various mutagenicity tests to a risk of heritable mutations in human beings? The available data give no direct information about mutagenic effects of chemicals on human germ cells in vivo. Application to humans rests on a series of assumptions about the response of human germ cells in relation to responses in other types of cells and in other species.

The analysis of the these test results does allow some generalizations about biologic extrapolation and about the nature of the available test database. On the first point, there is evidence suggesting that chemicals that test positive in some comparison tests for gene mutations or chromosomal aberrations have a likelihood of being positive in the SLT or HTT, respectively, but are not invariably so. To date, chemicals testing negative in the comparison tests have not produced clear positives in whole mammal germ-cell tests, but the database supporting this comparison is limited. Biologically, it seems unlikely that chemicals that are consistently negative in comparison tests would, in fact, be germ-cell mutagens in whole animals. But it is unlikely also that very many chemicals with negative results in one or two tests will have been tested in enough systems to allow an empirical test of that hypothesis.

Both L. Russell's and W. Russell's analyses described above suggest that a number of chemicals that test positive in the simpler comparison tests will be not be shown to cause mutations in spermatogonial germ cells. At present, however, it is impossible to know based on comparison test results, which chemicals are true human germ-cell mutagens and which are not.

A major limitation of the database is that nearly all tests have been in exposed male animals. Two strong animal mutagens—radiation and a chemical agent, ethylnitrosourea—do not appear to cause heritable mutations in the immature germ cells of female mice, but the data for females are not sufficient to draw firm conclusions.

The new technologies described in this report, which could be used to provide information on the types of mutations detected by the various tests, may improve our ability to apply the results of simpler tests to predicting human risk.

ATTEMPTS AT QUANTITATIVE EXTRAPOLATION

Not surprisingly, quantitative extrapolation is even less advanced than is qualitative extrapolation. However, there are some data bearing on quantitative relationships that may eventually be useful in predicting effects in human beings. First, there is some information about dose-response relationships. Second, some preliminary attempts are being made to determine relationships among certain corners of the parallelogram models described earlier in this chapter. One such effort is described below.

In an ongoing effort, a group of investigators is using a parallelogram approach to evaluate the effects of gamma radiation, cyclophosphamide (a medical drug) and benzo[a]pyrene (an environmental chemical) on mouse and human cells by assaying chromosomal aberrations and sister chromatid exchanges in mitogen-stimulated peripheral blood lymphocytes (PBLs). While these investigators have as yet little data, a paper presenting some preliminary results (170) lays out the rigorous procedures necessary for proper extrapolation of results from different studies of just one substance to predictions in untested systems. For instance, in the case of radiation, the authors relate an experimental result in irradiated cultured human PBLs to reports from the literature of the same chromosomal endpoint (dicentrics) in PBLs from patients who have received therapeutic radiation. Thus far, these authors have not presented any conclusions about the relationships they are studying.

Dose-Response Relationships

Extrapolation from high to low dose, and from high to low dose rate, requires knowledge of dose-response relationships. For heritable gene mutations, adequate data on dose-response relationships is limited to the effects of ionizing radiation and one chemical, ethylnitrosourea (ENU), in male mice. The available information comes from results of SLTs in mice, most of the work being done in a small number of laboratories in different parts of the world. Radiation and ENU have also been assayed in many short-term in vivo and in vitro tests in human somatic cells, so some generalizations may be drawn about the nature of dose-response relationships for these two agents. In both cases, there are independent effects of dose rate and of total dose. This means that a fixed dose may cause a different rate of mutations depending on the intensity of the dose, i.e., a more protracted administration may result in fewer mutations than if the total dose is administered at one time.

Radiation

Specific-Locus Test.—In a series of experiments from the mid-1950s to the present, a range of radiation doses, delivered at a range of dose rates, has been tested in male mice. Results are available for both spermatogonial and postspermatogonial stages. In spermatids and spermatozoa (postspermatogonial stages), the dose-response relationship is generally linear, and there is no effect of dose rate. This means that approximately the same mutation rate results from a short, high-dose exposure and from a chronic, low-dose exposure when the same total dose is given.

In spermatogonia, the early, pre-meiotic stage, for equal total exposure, radiation given at high dose rates causes more mutations per unit of dose than radiation at lower dose rates. At a high dose rate of 90 Roentgens per minute (R/min) or intermediate dose rate of 8 R/min (work cited in 111), the mutation rate decreases with decreasing dose faster than would be predicted by a linear relationship. At dose rates of 0.8 R/min or below, however, the rate of mutations per unit dose appears to be constant, and without a threshold. At low dose rates, the dose-response relationship is linear, and above about 0.8 R/min, the mutation rate per unit of dose rises more quickly than would be predicted by a linear relationship. Above a total dose of between 600 and 1,000 rem, the mutation rate begins to decline rapidly.

Investigators using the SLT offer an explanation for the observed dose-response patterns (111, 114,115). They postulate that the difference in dose-rate response between spermatogonia and postspermatogonial stages is a function of an active repair mechanism in metabolically active spermatogonial cells, which does not function at later stages. In the earlier, spermatogonial stages, a larger percentage of changes can be repaired before the spermatogonia complete meiosis when exposure is at a lower dose rate than is the case with an acute, high-dose-rate exposure. In post-spermatogonial stages when capacity for repair is low, the total radiation dose, irrespective of dose rate, is the determinant of the mutation rate.

Heritable Translocation Test.—Generoso and co-workers (135) have investigated heritable translocations induced by high-dose rate (96 R/min) irradiation of spermatogonial stem cells of mice. They report a linear dose-response relation between 0 and 600 R of total irradiation, and repetition of doses in this range gave additive effects up to 2,000 R. From these data, the expected increase in heritable translocations at high-dose-rates is calculated to be about 0.00004 per R.

Cytogenetics.—Waters and colleagues (170) describe a dose-response relationship for gamma radiation after both in vitro and in vivo irradiation, using as an endpoint a certain type of chromosomal mutation, dicentrics, in PBLs. The in vivo dose-response, which was derived from reports in the literature, is linear at lower dose rates, and quadratic at higher dose rates, meaning that the increase in the number of dicentrics rises faster than it would if the relation continued to be linear. The quadratic component may be explained by an interaction of mutational events causing some dicentrics, and the interactions being increasingly more likely as the dose rate increases.

Ethylnitrosourea (ENU)

Specific Locus Test.—As is the case with radiation, both total dose of ENU and dose-rate independently affect the mutation rate in mouse spermatogonia. Experimental data indicate that the response at doses below 100 mg/kg of body weight is "infralinear," meaning that as the dose is lowered, the mutation rate drops faster than would be predicted by a linear relationship. At doses between 100 and 400 mg/kg, the response appears to be linear (41). No threshold was detected over the range of doses tested, but the possibility of a threshold at a dose lower than 25 mg/kg (the lowest single dose tested) is not excluded by the data.

In an experiment examining mutational responses at different dose rates, the mutation rate was measured in mice that were given 10 weekly doses of 10 mg/kg of ENU, and compared with the mutation rate for a single dose of 100 mg/kg of ENU. The mutation rate for the fractionated dose was only about 15 percent of the rate for a single dose (112).

Russell (111) notes that, in light of information indicating that ENU reaches germ cells in doses proportional to injected amounts (see next section), these results cannot be explained by differences in metabolic processes. He interprets the infralinear portion of the dose-response curve and lower mutation rate that follows dose fractionation to be the result of effective mutational repair systems in spermatogonia, the same reasoning as in the case of radiation.

Unscheduled DNA Synthesis in Mouse Spermatids.—Carricarte and Sega (cited in 111) found the dose-response of "unscheduled DNA synthesis" in mouse spermatids to be linear over the range from 10 to 100 mg/kg, the same range over which W. Russell (111) found an infralinear relationship in the SLT. Sega (cited in 111) also measured "adduct formation" after injections of ENU, and found a linear response in the range from 5 to 100 mg/kg. One conclusion from these observations is that chemical interaction with DNA may not always be directly related to the rate of mutation, and this presents a major difficulty for Sobels' parallelogram, described earlier in this chapter.

Sister-Chromatid Exchange and Thioguanine Resistance.—Jones and colleagues (54) investigated the dose-response relationship for two somatic-cell endpoints in whole mice exposed to ENU. The frequency of sister-chromatid exchange (SCE) and the frequency of thioguanine resistant (TG^r) cells were measured in white blood cells in the spleen. Corresponding endpoints exist in human beings, which makes them potentially useful for extrapolating to human responses. The investigators measured both the response over time at several dose levels, and the response to a range of doses at several points in time after exposure.

Linear relationships were discovered between dose and reponse for both SCEs and TG^r cells, but the timing of these responses was quite different. For all dose levels, the highest SCE levels were measured on the first day after exposure, and SCE levels decreased back to baseline after about 70 days, whereas the TG^r response rose linearly over time to a peak after about 80 days for each dose level.

Generalizations About Dose-Response

While dose-response relationships cannot be generally and simply described some generalizations can be made. It appears that mutations in somatic cells are more predictable from the various experimental test systems available than are mutations germ-cell. Repair of mutations in germ cells may explain this difference. Both a better understanding of the mechanisms of mutation and repair at the molecular level, and empirical comparisons of test data on more substances will contribute to a better understanding of dose-response relationships. Once again, the new technologies may contribute to this understanding.

LIMITS OF EXTRAPOLATION

The main limitation to validating extrapolation models is a virtual lack of data to complete the parallelogram models. Until there is enough empirical information to determine whether qualitative and quantitative generalizations can be made, it is impossible to know how useful extrapolation could be. At present, *qualitative* extrapolation is at least a tentative guide for identification of mutagenic agents.

Even when data are available, many questions bedevil the use of experimental animal data to extrapolate risks to humans. In particular, little is known about the comparability of species with respect to activation, detoxification, and tissue distribution of specific chemicals, as well as other interspecies differences in metabolism. In addition, mutational responses of germ cells at different stages of development can differ for different types of mutagens. Many of these gaps in information could be filled by studies that are now technically feasible.

In using test results from somatic cell systems, there are differences in the sensitivities of various types of somatic cells and germ cells to different mutagens. Even within a single cell type, different gene loci can respond very differently to a specific mutagen. In making comparisons, for instance, between somatic and germ cells, if the same locus (or set of loci) cannot be tested, any differences in outcome cannot necessarily be attributed solely to a difference in cell type. These are a few of the many uncertainties in extrapolations from somatic to germ cells.

An organizational problem that affects extrapolation in a practical way is that collaborations have been rare among researchers working with different systems, and are particularly rare among scientists working in different laboratories. In addition, the test systems are not necessarily standardized among laboratories working with the same systems, so results from different laboratories cannot always be combined or directly compared.

Even relatively simple aspects of experimental procedures, for instance, the conditions under which cells or organisms are exposed to radiation or chemicals, vary enough among laboratories to render results incomparable to varying degrees.

Overall, the development of methods to extrapolate from different experimental test systems to the risk of heritable mutations in human beings is still in its infancy. While the various parallelogram models are appealing because of their potential usefulness, some of the data already available suggest that they may never broadly applicable. Both qualitative and quantitative improvements in the database of experimental results will be necessary before the usefulness of extrapolation can be reasonably assessed.

Chapter 8

Mutation Epidemiology

Contents

Chapter 8
Mutation Epidemiology

INTRODUCTION

This report is about biochemical and genetic techniques for studying heritable mutations. These new techniques will ultimately be used to study people who are suspected of being at high risk for excess heritable mutations. The types of epidemiologic activities in which mutation detection techniques may eventually be used are: surveillance, monitoring, and ad hoc studies. Before any technique can be used for those purposes, a series of validation studies will be needed, calling for different populations that are appropriate for study at different stages of development of the technologies.

Surveillance is a routine activity whose aim, in the context of this report, would be to measure the "baseline" rate of mutations in a defined population over the course of time, and to facilitate rapid recognition of changes in these rates. Following the distinction made by Hook and Cross (44), the term *monitoring* is reserved for observations over time in populations thought to be at increased risk for heritable mutations because of a known or suspected exposure to a known or suspected mutagen, for the purpose of helping the specific population in whatever way possible. *Ad hoc studies* of a variety of designs are carried out to *test hypotheses* about suspected causes of mutations.

The different aims of surveillance, monitoring, and ad hoc studies require that different criteria be applied for deciding when and whether to carry out one or more of those activities. Surveillance and monitoring are not designed as hypothesis-testing activities, though they may be sources for hypothesis development.

The reasons a population is chosen for *surveillance* may be largely opportunistic. It is unlikely that an entirely new system of data collection would be put in place for mutation surveillance. It is more probable that mutation surveillance would be added to an existing program that is established for another purpose, for instance, birth defects surveillance. The population covered must be large enough to generate reliable rates for mutational events that may be relatively rare, but there is no fixed requirement for size. Of the three types of activities, surveillance generally would involve the smallest effort and resource expenditure *per individual*, but because large numbers of people would be routinely subject to the surveillance test, the total cost could be large. There is, therefore, a great need to consider the costs and benefits of such a program before embarking on one, and for choosing the detection technique accordingly. The threshold for initiating mutation surveillance would be relatively high. Information about trends sometimes can be obtained by means other than full-scale surveillance, and some such studies for that purpose have been done.

The main reason for instituting a mutation *monitoring* program is concern about the potential effects of a mutagenic exposure in a specific group of people. If there is enough concern, there undoubtedly will be a greater expenditure of resources and effort per person than is the case for surveillance, meaning that more extensive contact with the population and testing would be justified. Given today's knowledge, it is hard to conceive of a situation in which there would be a concern only about heritable mutations, so any mutation monitoring effort would most likely be part of a larger program. The obvious concurrent concerns for exposures thought to be mutagenic are cancer and birth defects. Strict criteria based on tests of the statistical power to detect certain levels of effects are not appropriate for monitored populations, since the concern is about that particular group of people. Any finding is of interest in that situation. The information may be used incidentally to calculate upper limits of risk which could be generalized to other populations with similar exposures. The most important consideration in making a decision to monitor is that there is reasonable evidence suggesting that the population may be at a substantially increased risk.

This decision may well be influenced by political pressures to act, but ideally there should be a recognition that the best scientific judgment either does or does not support a monitoring effort.

The purpose of *ad hoc studies* is to test hypotheses about exposures and effects. This is the one place where it is imperative that studies be designed to achieve a high probability of detecting an effect if it is present. Such studies are valuable not only for the sake of the populations involved in the studies, but for their generalizability to other populations. Results of these studies can form the basis for public health actions, if levels of risk can be established.

VALIDATION OF MEASUREMENT TECHNIQUES

For the most part, the new technologies described in this report have not been "validated." They are new, and have not been applied to large numbers of people. Surveillance, monitoring, and ad hoc studies all require tools—in this case techniques for measuring mutations—with an acceptable degree of validity. That is, the tests must measure what they are designed to measure, within some definable limits. Generally, it is not possible to gather reliable information about a population and concurrently gather validating information about a technique used to measure outcome, unless another technique, with known validity, and known relationship to the new technique, is also applied in the study. Even though that is technically feasible, it is probably not an efficient way to gather validating data.

A first step in the validation process would be to use laboratory-prepared samples with known DNA sequences to confirm that the types of mutations that should theoretically be detected with a new technique actually are reliably detected. Beyond that stage, the need to move to clinical samples can be met using stored blood from individuals studied previously for other reasons. These stored samples need not be from parent/child triads initially, but triads will be needed at a later stage.

A number of research organizations are storing samples that would be appropriate for studies of mutations using new DNA techniques. The National Cancer Institute for example, is storing blood samples from cancer patients who have been treated with drugs and radiation, and the Radiation Effects Research Foundation has stored blood from Japanese citizens who were exposed to atomic bombs. In both of these cases, DNA is stored according to an established technique that uses Epstein-Barr virus to transform lymphocytes, thereby "immortalizing" the cells so they can be grown indefinitely. The transformed cells can be frozen for the long term in liquid nitrogen. Both sample preparation and long-term storage costs are substantial. With currently available technologies, it is unlikely that large numbers of samples will be stored. This limits the number and variety of samples available for validation studies, and also for later studies of people exposed to potential mutagens.

SURVEILLANCE AND DISEASE REGISTRIES

The methods and aims of disease surveillance have developed based on experience in infectious disease control. Reporting of vital statistics, in particular births and deaths for calculating birth and death rates, is also a form of surveillance. Surveillance of noninfectious diseases is a relatively recent development, with roots in the desire to track the incidence of cancer. Although the first national cancer surveillance system, which covers about 12 percent of the population, was put in place as recently as 1972, New York State instituted a reporting system for cancer cases in 1940, and Connecticut followed the next year. There are now dozens of cancer surveillance systems operating around the country and internationally, covering a range of populations from counties to whole countries (145).

Information about individual patients in cancer surveillance systems forms the basis for "cancer registries." The routine output of registries consists of cancer rates by sex, age, and race (where applicable) for each cancer site. Registries also are an important source for researchers investigating hypotheses about cancer causation. In this sense, cancer surveillance, with information about individuals recorded on registry forms, is similar to mortality statistics, with information about individuals recorded on death certificates.

It should come as no surprise that there are no "heritable mutation surveillance systems" now in place. There are, in various places around the world, registries of birth defects, which include records of at least some sentinel phenotypes. Beyond that, as this report shows, there are at present no techniques for detecting mutations that are suitable for use in a large-scale population surveillance program. Because developments have been so rapid, however, there may be one or several good candidate techniques within 5 or 10 years.

Surveillance traditionally has involved reporting, to a central place, information *already collected* by some segment of the health care system for reasons directly related to the health of individuals. Even for infectious diseases, only cases that come to the attention of physicians are reported. Active "case-finding" in the population is not a usual feature of surveillance. For chronic disease, the same is true. Cancers diagnosed by physicians are entered into registries. Case-finding programs, such as breast cancer screening programs, are instituted on the basis of their effectiveness in identifying cases early in the course of disease, for the benefit of the individual with the disease.

Infants are examined for birth defects because of the potential impact on the children and their parents' lives, and not mainly for the purpose of computing the rates of birth defects in a population. Testing programs for newborns, including biochemical tests for metabolic diseases (not necessarily a result of a mutation), also have been instituted because of their importance to the health of the individual. Nearly all States now require testing newborns for phenylketonuria (PKU), and some require additional tests. For instance, New York requires testing for PKU, sickle-cell anemia, and congenital hypothyroidism, which are moderately rare, and also for very rare conditions including maple syrup urine disease, homocystinuria, histidinemia, galactosemia, and adenosine deaminase deficiency (102). The tests do not impose an added burden on the newborn, since all are carried out using the same blood sample.

Cytogenetic techniques have not been used for population-based surveillance of chromosome abnormalities, but some large hospital-born series of newborns have been tested (102). Most of the recorded cases of chromosome abnormalities found in this way might eventually have been detected later in life because of health and reproductive problems, but some others might otherwise have gone undetected.

There is no formula for deciding whether to institute a surveillance program, but there are characteristics of the disease, of the population, and of the particular test to be used that contribute to the decision: 1) the seriousness of the disease (if the measured endpoint is known to be associated with a disease); 2) the ability to alter its clinical course after diagnosis; 3) the prevalence of the disease in the population; 4) the reliability and validity of the test; 5) the acceptability of the test to the population; 6) the cost of the screening program; and 7) the cost of *not* screening (i.e., the cost of treatment and social support). It is worth considering these factors in thinking about screening and surveillance for heritable mutations.

The idea of surveillance for heritable mutations represents a departure from the traditional applications of surveillance. It appears to be the case that most heritable mutations are not related to disease over the course of an individual's lifetime, and no predictions useful to the individual can currently be made about the effect of a heritable mutation in the absence of recognizable disease, beyond those that are associated with sentinel phenotypes and major chromosome abnormalities. As mutation detection techniques become more and more sensitive, in fact, a greater percentage of the mutations detected may not be related to a known effect on health.

Heritable mutation surveillance beyond reporting sentinel phenotypes will require more than just a reporting of events already detected. It will require imposing a test burden on a population for the sole purpose of collecting information about mutations that may never affect an individual's life. This argues against instituting surveillance. A reason in favor of surveillance is that it is clear that increased mutation rates *will* be looked for in special populations, those being monitored because of worries that they have been exposed to a mutagen. Surveillance systems can provide a range of estimates of "baseline" or "background" rates, even though they may be from different populations. In a more general sense, one of the original aims of surveillance is relevant: to substantiate long-term trends and patterns in health events and to detect changes that may be addressed by public health action.

MONITORING AND EXPOSURE REGISTRIES

Monitoring is the "long-range observation of individuals who are at presumptive high risk for adverse outcomes because of specific life events," (44) in particular, exposures to suspected mutagens. The event may be catastrophic, such as exposure at the time of detonation of an atomic bomb, or a chemical plant explosion. Or the "event" may be long term, such as an occupational or an environmental exposure. There are about two dozen populations around the world currently monitored for long-term health effects, and some of those programs include various studies of heritable mutations.

The most extensive population monitoring, including monitoring for mutations, is of the Japanese residents of Hiroshima and Nagasaki, many of whom were exposed to substantial amounts of radiation during World War II when atomic bombs were detonated in those cities. The population around a chemical plant that exploded near Seveso, Italy in 1976, releasing several pounds of dioxin, is the subject of health monitoring activities, including monitoring for birth defects. The people exposed to methyl isocyanate in Bhopal, India, will undoubtedly be followed for years to come. Because these groups were exposed, and because it is conceivable that something could be done to alleviate health problems if they are detected early, or if warning signals are picked up, they are being monitored; the scientific knowledge gained as a result is a secondary benefit.

The most prominent examples of chronic exposures are from occupational activities and toxic chemicals in the environment. The populations exposed to hazards often are not geographically determined, but may be a collection of workers from around the country. Workers exposed to radiation in the nuclear power industry are an example of this. There are several "exposure registries" in existence worldwide, though none specifically because of a perceived increased risk of mutations. One such registry has the names of all workers who were exposed to dioxin during the manufacture of various chemicals in this country. There also is an international dioxin registry, with names of workers from all around the world. The registry does not, however, have information about the health status of those workers. A similar registry for workers exposed to beryllium exists in this country. A report prepared for the Nuclear Regulatory Commission in 1980 recommended that a registry be started for workers exposed to low-level ionizing radiation in certain types of workplaces, because of a possible increased cancer risk (29). These registries could be used for monitoring and as a potential population to include in ad hoc studies.

AD HOC EPIDEMIOLOGIC STUDIES

Surveillance, monitoring, medical case reports, and laboratory research can all lead to hypotheses about possible causes of heritable mutations. An investigator wishing to test a hypothesis must find suitable subjects to study, in contrast to a monitoring activity, where the existence of the exposed population is the reason for acting. A study should be undertaken only if there is a good chance of answering the question of interest. Disease and exposure registries are common sources of individuals to study, depending on the question.

A cohort design will probably prove the most useful approach for studies of heritable mutations, though case-control studies of sentinel phenotypes may also prove valuable. A cohort study involves identifying a group of individuals, some exposed to the suspected mutagen and some not exposed. The health outcomes, i.e., the presence or absence of mutations in offspring, of the two sub-cohorts are compared. A higher rate of mutations in the exposed group would signify an "association" between the exposure and heritable mutations. Statistical tests are applied to the results to estimate the likelihood of the result occurring if in fact there was no real difference in mutation rates between the two groups.

In a case-control design, a group of "cases," individuals with conditions of interest, e.g., sentinel phenotypes, is compared to a group of individuals who do not have the condition of interest, but who are otherwise similar demographically. The cases and controls are compared according to their past histories of exposures or other characteristics that might be associated with the mutation and an assessment made as to whether their histories differ in important ways.

The important question for all studies is not just whether the exposure is "associated with" mutations, but whether it *causes* them. That is a difficult if nearly impossible judgment to make in most instances, but there are some generally accepted guidelines for evaluating the likelihood of an association being causal based on epidemiologic evidence. These are:

1. **Consistency:** The association is observed in studies by different investigators, at different times and in different populations, and in studies of different designs.
2. **Strength:** The size of the effect of an exposure is the measure of strength of association. This is usually measured as an estimate of relative risk (a ratio of the rate of mutations in an exposed group to the rate in an unexposed group). The presence of a dose-response relationship, that is, the size of the effect changes in a logical way with the level of exposure and in at least some cases, with the dose rate.
3. **Specificity:** Specificity refers to the degree to which the exposure is associated exclusively with the outcome of interest, in this case a mutation, and the degree to which a mutation is associated exclusively with the exposure. The concept of specificity derives from study of infectious disease and is relevant to the study of mutations (and chronic diseases generally) only in special cases, for example, a specific mutation that almost never occurs in the general population but appears to be exclusively related to a particular exposure. While a highly specific relationship can provide positive evidence for a causal relationship, a lesser degree of specificity does not necessarily argue strongly against causality.
4. **Temporal Relationship:** The exposure must occur before the effect. In the case of heritable mutations, the picture is more complicated. See chapter 6 for a discussion of the timing of exposure for males and females for a plausible effect on germ cells.
5. **Coherence:** All available information from medical and biological science, and from epidemiologic observations and studies, fits together in a way that supports the hypothesis. The greater the variety of information, and types of study designs, the stronger the finding of coherence.

These criteria are quite stringent, and even in the best of cases, often cannot be met, but they are useful as standards.

POPULATIONS TO STUDY

There are elements in the environment that damage human health under certain conditions of exposure. Biologic, chemical, and physical agents cause acute and chronic diseases in humans. At present, there are no exposures unequivocally known to cause heritable mutations in human beings. A combination of factors, including the rather insensitive methods for detecting heritable mutations that have been available, and the possibility that human germ cells may not be very susceptible to some mutagens, probably contribute to this situation. As a consequence, investigators looking for the effects of mutagens must do so in people who have been highly exposed to agents that are very likely to be mutagenic in germ cells. There are not very many large groups of people fitting that description, a fact that many might find surprising.

Radiation-Exposed Groups

Radiation causes heritable mutations in laboratory mice and is the most likely potential germ-cell mutagen to which large numbers of human beings have been exposed, either intentionally or accidentally. The largest population with a known high radiation exposure, the Japanese atomic bomb survivors, continue to be followed for effects on cancer incidence, birth outcomes, and heritable mutations. Heritable mutations have been studied by clinical observations, cytogenetic techniques, one-dimensional electrophoresis of blood proteins, and more recently with the most sensitive technique of two-dimensional gel electrophoresis of blood proteins (see ch. 3).

A report was prepared in 1980, under contract to the Nuclear Regulatory Commission (NRC), that evaluated opportunities for studying the health effects of low-level ionizing radiation (29). The report is focused on cancer, but the evaluation methods apply equally to studying mutations. The authors initially identified 100 candidate populations. About 30 remained after two broad criteria were applied: 1) that there be data identifying exposed individuals, and 2) that there were at least 10,000 people in a single population group or one comprising several similar groups. Those 30 populations were evaluated further, and recommendations made that if additional studies were to be undertaken, three occupational groups and one group with environmental exposure held out the greatest promise of yielding a reliable result. Even the best of these, however, has a relatively low power: less than a 50 percent chance of finding an excess of cancer if it exists. In general, this level of power would be unacceptable in an epidemiologic study. Although political considerations might influence a decision to go ahead with a study, they do nothing to increase the power of the method.

The power figures for these studies refer to cancer detection, and the probability of detecting heritable mutations is undoubtedly far lower, making it unlikely at best that anything could be learned about radiation and heritable mutations by studying any of these groups with currently-available methods.

The report to the NRC contained one other recommendation, that a registry for radiation workers be initiated. The registry would maintain information about radiation doses and some information about other exposures. This recommendation has not been acted on. There are examples of radiation-exposure registries, but these are mainly for people acutely exposed accidentally, and not for the more usual long-term chronic exposures of workers.

Cancer Patients

Treatment for many cancers includes chemotherapy with cytotoxic drugs, some of which are carcinogenic in laboratory animals and mutagenic in vitro, and treatment with high doses of radiation. There is a growing body of evidence that cancer patients are at a severalfold increased risk of developing second cancers, and some of these second cancers may be attributable to treatment of the first cancer with drugs and radiation (see, e.g., ref. 149). Cancer is mostly a disease of old age, but certain cancers have their peak incidence in younger people. Hodgkin's disease, for instance, occurs with greatest frequency in young men. Childhood leukemias, some brain cancers, and tumors with strong genetic components, e.g. Wilms' tumor, retinoblastoma, and neuroblastoma, occur in the first few years of life. As treatment for these early cancers has improved over the last two decades, larger numbers of people are surviving, and it is these survivors who are at an increased risk of a second cancer, and possibly of heritable mutations.

Results from four studies of the offspring of childhood cancer survivors, and nine studies of offspring of adult cancer patients have been published as of mid-1985. Several other studies are in progress (82).

The combined published studies represent more than 700 cancer patients (both male and female) and more than 1,500 pregnancies, about 1,200 of which resulted in live births. Four percent of the liveborn babies had major birth defects, which is similar to the incidence in the general population. Two of the liveborn children had cancer. One had a hereditary bilateral retinoblastoma, as his father had. The other, the daughter of a brain cancer survivor, had acute myelocytic leukemia. One child had a condition that could have been the result of a new mutation, the Marfan syndrome, which fits the definition of a sentinel phenotype. Several other children had defects that might have had genetic components, but none of these represented sentinel phenotypes.

The largest study of offspring of childhood cancer survivors, including about 2,300 individuals from five population-based cancer registries, is nearing completion. Preliminary results indicate no increased risk of cancer in offspring compared with a control group, but the analysis is not yet final (82). Another long-term followup study, with more than 3,300 cases enrolled to date, is under way in the United Kingdom. No results are yet available from that study (82).

The findings of a large international cooperative study of second tumors in children treated for cancers are provocative (142). Overall, 12 percent of children who survive at least 2 years after a first cancer develop a second cancer sometime during the 25 years following the first cancer. Most of the patients in the study were treated with high-dose radiation therapy. The risk of second cancers was highest among children with cancers known to be strongly genetically influenced. In that group, there may well be a genetic defect that predisposes to mutations, e.g., a faulty repair mechanism, which could also be related to a higher risk of heritable mutations in that group.

Cancer registries are the most numerous registries of any type, and cohorts of treated patients and patients with second tumors are relatively easy to identify, compared with identifying other populations potentially exposed to mutagens. These groups should be considered when studies of heritable mutations using the new technologies become feasible.

Other Populations

A study of birth outcomes in people who had attempted suicide by self-poisoning in Hungary is an example of opportunistic use of available information (23). A cohort of about 1,300 individuals who took large doses of drugs in suicide attempts has been studied since 1976. Early on, the investigators looked for short-term effects on somatic cells, using cytogenetic and biochemical testing. Long-term followup of birth outcomes examined spontaneous abortions, ectopic pregnancies, stillbirths, low birthweight, and congenital anomalies. The study suffered a large loss of followup of study subjects, but in those evaluated, no important excesses in any of these endpoints were discovered.

CONCLUSIONS

A very important question is answered by simply observing birth outcomes in people thought to be at high risk, namely whether those individuals are at risk of having children with serious diseases and disabilities. The new mutation detection technologies discussed in this report may greatly increase the power to identify mutations in studies such as those described above, adding another dimension to knowledge about the relationship between exposure to mutagens, the presence of detectable mutations in DNA, and the existence of observable health effects.

Chapter 9

Mutagens: Regulatory Considerations

Contents

Chapter 9

Mutagens: Regulatory Considerations

INTRODUCTION

It is clear from the information gathered in this report that we are some years distant from being able to muster convincing direct evidence of any but very large increases in the rates of heritable mutations in human beings. The ability to predict from experimental data which agents are likely to increase the mutation rate *if* human beings were exposed has not been put to the test. Even for the most likely mutagens, the ability to make quantitative extrapolations is relatively undeveloped. Nonetheless, it is reasonable and prudent to accept that the environment may contain human germ-cell mutagens and that, to the extent possible, human beings should be protected from them at levels that might cause heritable mutations.

Ionizing radiation was recognized as a cause of heritable mutations in fruitflies in the 1920s, and the Federal Government has since made efforts to protect workers and the public from excessive radiation exposure. Widespread concern about the mutagenic potential of chemicals is more recent, an issue brought into focus by the environmental movement that took shape in the late 1960s. While the potential cancer-causing properties of man-made chemicals have been the driving force behind environmental health laws, two more recent laws specifically mention mutation as an endpoint against which the public should be protected. About a dozen other statutes include language broad enough to charge the Federal Government with the responsibility to protect against heritable mutations. Evidence from current methods for measuring mutation rates suggests that basing regulation of environmental agents on carcinogenicity will likely assure protection against heritable mutations, but new, more sensitive detection technologies, such as those discussed in this report, may necessitate a reexamination of that conclusion.

FEDERAL INVOLVEMENT IN PROTECTING AGAINST GENETIC RISK

Radiation Protection

In 1928, the newly created International X-Ray and Radium Protection Commission was charged by the Second International Congress on Radiology with developing recommendations for protection against radiation (156). The following year, the Advisory Committee on X-Ray and Radium Protection was formed to represent the U.S. viewpoint to the international commission. These two bodies were the forerunners of the current International Commission on Radiological Protection (ICRP) and the National Council on Radiation Protection and Measurements (NCRP), the latter chartered by the U.S. Congress in 1964. The ICRP and NCRP have, since their first recommendations in the 1930s, based their acceptable radiation exposure limits on both heritable and somatic effects. The limits recommended have been lowered over the years, reflecting increased knowledge about radiation effects, and particularly about the effects on the population of low levels of radiation.

Neither the ICRP nor NCRP recommendations have the force of law, but by and large, they have formed the basis for the radiation protection limits adopted by U.S. regulatory agencies. The first Federal entity officially charged with providing the agencies with guidance for developing radiation protection standards was the Federal Radiation Council (FRC), established in 1959. In 1960, FRC issued recommendations for both occupational exposure and exposure of members of the public, which drew on ICRP and NCRP work (156). Over the years, the National Academy of Sciences Committee on the Biological Effects of Ionizing Radiation and the United Nations Scientific Committee on the Effects of Atomic Radiation have also been influential in providing analyses that undergird exposure limits.

The National Environmental Protection Act of 1970 transferred the responsibilities of FRC to the new Environmental Protection Agency (EPA). EPA administers several environmental health statutes under which that Agency is responsible for setting standards for radiation exposure in specific conditions. Under its broader responsibilities, EPA has provided guidance for exposure from diagnostic X-rays, which is the regulatory responsibility of the Federal Food and Drug Administration (FDA), and for exposure of uranium miners, who are the responsibility of the Mine Safety and Health Administration (MSHA).

Agencies other than EPA with responsibility for some aspects of radiation protection include: the Nuclear Regulatory Commission, the Department of Energy, the Department of Defense, FDA, MSHA, the Occupational Safety and Health Administration (OSHA), and the Department of Transportation. The States have responsibilities as well. Each entity, depending on its specific charge, is required to protect workers, the public, or both in accordance with the guidance provided by EPA.

The basic occupational and population exposure guidelines have not been revised since 1960. In 1981, EPA proposed new occupational guidelines (153), in line with 1977 ICRP recommendations (47), but these have not been made final, and they do not represent a change in total acceptable dose from the earlier guidelines. They do, however, place less of the emphasis on mutagenesis, and relatively more on somatic effects than do the 1960 guidelines. The ICRP includes in its risk estimates only genetic effects occurring in the first two generations after exposure. That probably accounts for roughly half of the total genetic effect, whatever its size.

The current occupational exposure limit is 5 rem total body dose per year, with not more than 3 rem total body dose from occupational exposure in any one quarter of the year, and a more detailed breakdown for different groups of organs. The 1977 ICRP recommendations abandon the specifications by organ, and use a weighted whole-body dose.

Today, almost all radiation exposures of U.S. workers are well below the regulated limits, though there are exceptions. The quantitative limits set by the ICRP and NCRP and adopted by Federal groups, are accompanied by the "ALARA principle"—that radiation exposures should be "as low as reasonably achievable."

Exposure to the public are to be limited to below 25 millirem (mrem) to the whole body, 75 mrem to the thyroid, and 25 mrem to any other organ. These levels mainly affect the regulation of radionuclides in air and the disposal of radioactive waste. The numbers come from ICRP and NCRP recommendations, and are based on consideration of both genetic and somatic effects.

Agents Other Than Radiation

Congress formally recognized the need to protect against chemical mutagens in the Toxic Substances Control Act of 1976 (TSCA), and again in 1980 in the Comprehensive Environmental Response, Compensation, and Liability Act (CERCLA, or "Superfund"). These two laws are administered by EPA, as are other statutes that include broad mandates to protect the public from environmental hazards. Other laws designed to protect citizens from external agents under which chemical mutagens could be regulated include the Federal Food, Drug, and Cosmetics Act, administered by FDA; the Occupational Safety and Health Act, administered by OSHA; the Consumer Product Safety Act, administered by the Consumer Product Safety Commission; and the Atomic Energy Act, through which the Nuclear Regulatory Commission is empowered to protect certain workers from radiation hazards.

Although it is almost certain that chemicals that might cause heritable mutations have been regulated, no regulations have been written or standards set for these agents because of that property. In a few cases, the mutagenic potential of chemicals has been considered by regulatory agencies, but carcinogenic properties have driven standard-setting.

OSHA has included thorough reviews of mutagenicity data in notices of regulatory actions for two of the best-publicized chemical hazards of the 1980s: ethylene oxide (EtO) and ethylene dibromide (EDB). Tests for heritable mutagenicity in Drosophila and experimental mammals have yielded positive results in at least some systems for both of these chemicals. While results of mutagenicity tests are included in the Federal Register notices for these chemicals, quantitative extrapolations for both the final EtO standard (152), and the proposed rulemaking for EDB (151), are based on protecting against carcinogenicity.

EPA has recognized germ-cell mutagenicity as a class of adverse effects, particularly in its responsibilities under the Federal Insecticide, Fungicide, and Rodenticide Act, under which it, in addition to OSHA, has acted to regulate exposures to EtO and EDB. Unique among the regulatory agencies, EPA's Reproductive Effects Assessment Group (in the Office of Research and Development) has prepared guidelines for mutagenicity testing, which are described later in this chapter.

REGULATORY ISSUES

Currently there does not appear to be a scientific basis for the specifics of regulatory action against mutagens. The following questions face regulators and the scientific community involved in mutation research:

1. What is an appropriate regulatory definition of a probable human germ-cell mutagen and how does that definition relate to what is known about mutagens from epidemiologic studies and experimental studies?
2. Is it possible to derive quantitative estimates of the risk of heritable mutations in humans from experimental evidence in animals or from somatic-cell mutation tests in human beings? If it is not, what kinds of information are necessary before such extrapolation is possible?
3. How likely is it that a substance will require a more stringent standard as a mutagen than it will as a carcinogen or for other toxic effects?
4. How will information from the new technologies for detecting heritable mutations that are described in this assessment change our perception of the kinds of "adverse effects" against which regulation should be directed?

The discussion in the remainder of this chapter addresses these questions.

A Regulatory Definition of a Germ-Cell Mutagen

Strategies for regulating mutagens to protect public health cannot today rely on data from current studies of heritable mutations in human beings. Just as is the case in regulating carcinogens, a regulatory definition must serve as a substitute, particularly for making judgments about the potential risks of new substances.

Certain lessons can be learned from experience in regulating carcinogens. (For a discussion of the issues surrounding carcinogen regulation, see 145.) The most convincing evidence for carcinogenicity, from well-conducted epidemiologic studies, is that human beings have developed cancer after exposure to a given agent. If an increase in *genetic disease* could be convincingly shown to be related to a specific agent, there would certainly be no problem in acting against that agent. The spirit of the regulatory laws, however, embody the concept of taking protective action before people are harmed. In the absence of direct evidence from human beings, regulators must rely on indirect evidence from a variety of experimental test systems which *will never be absolutely predictive* of effects in human beings. A regulatory definition of a mutagen will be pragmatic and rely on information that it is possible to collect, and on a number of untested assumptions.

EPA is the first U.S. regulatory agency to have proposed "Guidelines for Mutagenicity Risk Assessment" (154). The guidelines require evidence of: 1) mutagenic activity and 2) chemical interactions in the mammalian gonad. The decision to regulate is to be based on a "weight-of-evidence" determination. EPA has proposed no formal method for quantitative extrapolation by which acceptable exposure levels could be set.

According to the EPA guidelines, evidence for mutagenic activity may come from tests that detect point mutations and structural or numerical chromosome aberrations. Structural aberrations include deficiencies, duplications, inversions, and translocations. In the absence of evidence of heritable mutations in human beings, evidence from a variety of experimental test systems may be invoked. For mutagens that cause point mutations, whole animals tests (e.g., the mouse specific-locus test) provide the highest degree of evidence, but these tests are relatively more expensive than short-term tests, and there is a limited capacity for laboratories to perform them. Other tests for point mutations include those in bacteria, eukaryotic micro-organisms, higher plants, insects, and mammalian somatic cells.

Structural chromosome aberrations can be detected either in somatic or germ cells in different assays. The organisms used include higher plants, insects, fish, birds, and several species of mammals. Mutagens that cause numerical changes in chromosomes may be missed by the tests that directly measure DNA damage. Tests specifically directed at detecting changes in chromosome number are not as well developed as are those for point mutations or structural changes in chromosomes. Tests are in various stages of development in fungi, Drosophila, mammalian cells in culture, and intact mammals, including mammalian germ-cells tests.

Results from tests that measure endpoints other than mutagenicity directly may also be used in judging the potential mutagenicity of a substance. DNA damage, unscheduled DNA synthesis in mammalian somatic and germ cells, mitotic recombination and gene conversion in yeast, and sister chromatid exchange in mammalian somatic and germ cells are cited by EPA as tests that provide evidence known to be correlated with mutagenicity, though they measure other genetic events.

Evidence from various kinds of mutagenicity tests is weighted with regard to the relationship of the test to human germ-cell mutation. Greater weight will be given to results from tests in: 1) germ cells over somatic cells, 2) mammalian cells over submammalian cells, and 3) eukaryotic cells over prokaryotic cells.

EPA lists two classes of evidence for chemical interactions in the mammalian gonad: sufficient and suggestive. Sufficient evidence is from studies in whole mammals that demonstrate, for example, unscheduled DNA synthesis, sister chromatid exchange, or chromosomal aberrations in germ cells. Adverse effects on the gonads or on reproductive outcomes after exposure, which are consistent with the substance reaching the gonads but which do not indicate direct interaction with DNA, are considered as providing suggestive evidence.

The final step in EPA's mutagenicity risk assessment is the weight-of-evidence determination, which classifies the evidence for potential germ-cell mutagenicity as "sufficient," "suggestive," or "limited." In this step, results of tests plus any information about effects in human beings is evaluated. Sufficient evidence consists of a positive mammalian germ-cell test. In addition, positive responses in at least two different test systems, at least one of which is in mammalian cells, and evidence of germ-cell interaction, together constitute sufficient evidence. Evidence of lesser quantity and/or quality of both mutagenic response and germ-cell activity constitute suggestive evidence. Limited evidence consists of positive results in either mutagenicity assays or tests for chemical interactions in the gonad, but not both.

EPA's guidelines became final in September 1985. Currently and for the foreseeable future, the greatest value of EPA's guidelines is the recognition of germ-cell mutagenicity as a legitimate endpoint to consider in assessing the potential adverse effects of substances in the environment.

Quantitative Extrapolation

If the levels of risk from suspected human germ-cell mutagens is to be estimated in the absence of direct evidence of harm in human beings, data from experimental systems must be used in a "quantitative extrapolation." The experimental systems are basically those mentioned in the EPA guidelines discussed in the previous section, and those discussed elsewhere in this report. The three categories of tests are: 1) whole animal heritable mutation studies; 2) animal somatic-cell mutation studies, either in vivo or in vitro; and 3) human somatic-cell mutation studies, either in vivo or in vitro. Unfortunately, the kind of information (i.e., measures of human mutations) that would link results from these three categories of tests to human heritable mutations is scanty. It is encouraging, however, that using tests available now, such information can be generated at least for some substances. For EtO and EDB, for instance, mutagenicity data are available in both somatic and germ-cell systems in animals, and some somatic cell (cytogenetic) data are available from human beings exposed at known occupational levels.

Obtaining more information to fill in the arrows of the extrapolation "parallelograms" presented in chapter 7 of this report should be a high priority for regulators. In fact, EPA's Reproductive Effects Assessment Group has collaborated with other groups in the Federal Government to fund such studies (168). Without the kind of information that would come from coordinated studies in several test systems, there is little chance of writing a successful regulation that limits exposure to a specific level (short of a complete ban for an agent acknowledged to be unacceptably risky at any level). Given the experience with carcinogens, a regulation that states an exposure level without adequate experimental evidence and theory behind it will not survive a court challenge, which, in the United States today, appears to be the final test of a regulation.

Even with good experimental data, some of the same problems that plague extrapolating from animals to humans to determine acceptable exposure levels for carcinogens are certain to hinder quantitative extrapolation for estimating levels of mutagenic risk at specific levels of exposure. In carcinogen extrapolation, there still are unresolved controversies about the appropriate conversion factors between species and about the shape of dose-response curves. The latter is important because most animal bioassays test extremely high doses in relation to the animals' body weights, while humans are generally exposed at lower levels over longer periods of time. Though the details of extrapolation for mutagenicity differ from those for carcinogenicity, the problems will undoubtedly be similar. Right now, there is not enough information about the relationships between results in various test systems to address intelligently the practical problems of actually performing extrapolations.

Given the appropriate information, it may become possible to carry out quantitative extrapolations for mutagenicity. When the time comes, the regulatory agencies will need to require that the appropriate tests are done, either by manufacturers or by the Federal Government. The testing requirement may take various forms, which may vary by statute.

Mutagenicity and Carcinogenicity

As mentioned previously, while the regulatory apparatus exists for acting against human germ-cell mutagens, in fact no regulations based on germ-cell mutagenesis exist except for radiation exposure limits. There are several reasons for this. First, there are no proven human germ-cell mutagens, and only a limited number of presumptive human mutagens known from animal tests. Second, it has often been thought that regulations based on demonstrated carcinogenicity would automatically protect against mutagenicity as well. In fact, however, this may not always be true. Voytek (167) reported a preliminary assessment indicating that the risk of heritable genetic disease in the first generation after exposure to EDB was greater than the lifetime risk of cancer in the exposed individuals, based on an extrapolation from animal data.

It is widely held that a somatic mutation is a necessary step in the development of cancer. Many substances that are mutagenic in bacteria, in cells in culture, and in Drosophila also are carcinogenic in laboratory rats, mice, or both. Short-term tests based on mutagenicity in these lower organisms are therefore used as screens for carcinogenicity. The most widely used screening test for carcinogenicity is the Ames test, which measures mutagenicity in strains of the bacterium Salmonella. Under some statutes (e.g., TSCA), negative results in short-term tests, meaning that the substance is not mutagenic in those systems, can obviate the need for a long-term bioassay, a savings of up to $1 million to a manufacturer (145). Positive results in short-term tests, meaning that the substance is mutagenic in these systems, have not been accepted as a basis for regulation under any statute, but they have probably halted the development of new chemicals. Manufacturers know that positive mutagenicity tests may trigger the requirement for a long-term bioassay, which in a high proportion of cases will turn out positive. The product, whatever it is, might never get to market. Rather than risk a financial loss, many manufacturers simply will not proceed with that product.

New Methods for Measuring Mutation Rates and Their Potential Effects on Regulation

The new techniques for detecting heritable mutations and the somatic-cell techniques that eventually may be used to predict germ-cell mutagenesis will lead to the consideration of effects that are increasingly removed from measurable or even hypothetical adverse health effects in humans. Mutagenicity endpoints may be detected that could be more sensitive than those currently used to predict carcinogenicity. In the regulatory context, judgments will have to be made, most likely in the absence of certain knowledge of effects, about the appropriate actions to be triggered by demonstrations of various kinds of changes in DNA, detectable by various analytic methods, that can be convincingly linked to specific exposures. From a public health standpoint, it is most appropriate to act under the assumption that mutations of any kind are deleterious, and that environmental agents at levels that cause any reliably detected changes in the DNA should be subject to available regulatory controls. This does not get around the problem of defining "safe" or "acceptable" exposure levels, however. Animal experiments will be needed to explore the quantitative relation between subtle changes detectable anywhere in DNA and the levels of adverse effects that might be observed in the animal.

At present, testing for safety is a significant part of the research and development investment in new products, whether they are chemicals, drugs, or food additives; determining the risks of substances already in the environment is a significant task for the Federal Government. If they become available, new mutagenicity testing technologies using experimental animals could either impose significant new testing requirements in addition to those already in place, or could replace some expensive and not entirely reliable tests for carcinogens. Regardless of the methods used to detect mutations, the relation between specific types of changes in DNA and health effects will have to be studied experimentally to shed light on the meaning of a "positive result." Until this knowledge is available, the ability to detect an effect without knowing the likelihood of any health consequences for human beings will remain a thorny public policy question which scientists, regulators, and politicians must address.

Appendixes

Appendix A

Federal Spending for Mutation Research

A quick review of the origins of biotechnology illustrates the difficulty of identifying the specific areas of research that would further our knowledge of human mutations, and, in turn, the problems of estimating the current amount of Federal expenditures for such research.

In 1968 and 1969, Hamilton O. Smith, a microbiologist at Johns Hopkins University, had no thought of studying human heritable mutations. No venture capital firm was looking for biotechnology companies to invest in, and had it been, it would have been disappointed. No such company existed.

Smith was studying how some bacteria can take up molecules of DNA from their growth medium and recombine it into their genetic material. He knew that the bacteria would not take up DNA from other species of bacteria or from viruses, and he added some viral DNA to the experiment expecting that it would remain inert and intact while the bacterial DNA underwent recombination. Instead, the viral DNA was rapidly degraded. Looking at that result, Smith hypothesized that a bacterial enzyme(s) could discriminate between foreign DNA (from the virus), which it degraded and self DNA (from the same kind of bacteria), which it left alone. He purified the enzyme and showed that it recognized particular sequences of nucleotides on the viral DNA and cut it at those sites (128).

These experiments, which identified and characterized the first known restriction enzyme, opened the door to the discovery of hundreds of others. The use of restriction enzymes, allowing precise cutting of DNA and precise joining of DNA pieces from different parts of an organism's genome as well as joining of pieces from different organisms, led to fundamental advances in the basic approach of research in molecular biology and genetics (146,147). Accordingly, these enzymes are basic ingredients of the new technologies for studying human heritable mutations described in this report.

However, someone in 1969 who had tried to identify research projects that would contribute to our understanding of human mutation rates would almost certainly not have included Smith's. His was basic research directed at understanding the mechanism of recombination in an organism far removed from humans. This OTA report provides additional examples of the difficulty of estimating how much money is being spent on research that may be important to understanding human mutations.

Current Federal Expenditures for Mutation Research

OTA queried Federal research and regulatory agencies about their support of research directed at understanding human mutations. Questions were asked about the amount of money spent specifically on human mutation research and the amount spent on "related" areas of biological research. The first category was tightly defined—the research had to focus on development or applications of methods for detecting and/or counting human somatic or heritable mutations *in vivo*. The category of "related" research was much broader, including studies examining genetic contributions to such common diseases as diabetes and arthritis, studies of mutagenesis and repair mechanisms, development of instruments for measuring cellular effects of various kinds, and tests for mutagenic activity in lower organisms as well as in cultured human cells.

The expenditures shown in table A-1 must be treated with some reservation about their accuracy, the total of $14.3 million estimated as the amount spent on human mutation research being more precise than the approximately $207 million estimated for related genetic research. The $14.3 million may be an overestimate—

Table A-1.—Federal Expenditures for Research Related to Human Mutations

Federal agency	Human mutation research	Related genetic research
Department of Energy	$ 6,220,000[a]	$ 30,123,000
Department of Health and Human Services:		
Centers for Disease Control	277,000	1,387,000
National Institutes of Health	7,244,000[b]	156,959,000[c]
Food and Drug Administration		1,200,000
Environmental Protection Agency	606,000	1,189,000
National Science Foundation	0	16,336,000
Total	14,347,000	207,194,000

[a]This includes $1,850,000 (or half of the total $3,700,000 at current exchange rates) spent annually for the U.S. contribution to the Radiation Effects Research Foundation, the group that coordinates ongoing genetic studies of the survivors of the Nagasaki and Hiroshima bombings and their offspring.

[b]This total was obtained by inspecting grant titles and reading grant abstracts to identify those that focused on human mutation research.

[c]This figure is the budget for the Genetics Program in the National Institute of General Medical Sciences. It does not include all genetics research at NIH, some of which goes on in other institutes.

SOURCE: Office of Technology Assessment.

some projects studying clinical effects of mutations may provide no information about detection—and the $207 million for related genetic research is probably an underestimate, since much genetic research is supported by parts of the National Institutes of Health (NIH) not queried. For comparison, the total NIH budget for the fiscal year 1985 was about $4 billion. Most of the NIH funding for mutation research is directed at identifying mutations in at-risk families (where specific tests for specific mutations are appropriate), whereas DOE supports a search for more general methods to detect mutations.

Successes of the research efforts in human mutation research and related genetic research will probably generate a requirement for additional funding in the near future. It will be a costly venture to take any new technique (such as those described in ch. 4) and apply it to studies of populations sufficiently large to provide a good possibility of yielding information about human mutations. For instance, the ongoing monitoring of the Nagasaki and Hiroshima populations using clinical examinations and protein analyses (see ch. 3) costs about $3.7 million annually to the U.S. and Japanese governments together (48).

Appendix B

Acknowledgments and Health Program Advisory Committee

HEALTH PROGRAM ADVISORY COMMITTEE

Sidney S. Lee, *Chair*
President, Milbank Memorial Fund

H. David Banta
Project Director
STG Project on Future Health Technology
The Netherlands

Rashi Fein
Professor
Department of Social Medicine and Health Policy
Harvard Medical School

Harvey Fineberg
Dean
School of Public Health
Harvard University

Patricia King
Professor
Georgetown Law Center

Joyce C. Lashof
Dean
School of Public Health
University of California, Berkeley

Alexander Leaf
Professor of Medicine
Harvard Medical School
Massachusetts General Hospital

Frederick Mosteller
Professor and Chair
Department of Health Policy and Management
School of Public Health
Harvard University

Norton Nelson
Professor
Department of Environmental Medicine
New York University Medical School

Robert Oseasohn
Associate Dean
School of Public Health
University of Texas

Nora Piore
Senior Fellow and Advisor to the President
United Hospital Fund of New York

Dorothy Rice
Regents Lecturer
Department of Social and Behavioral Sciences
School of Nursing
University of California, San Francisco

Richard Riegelman
Associate Professor
George Washington University
School of Medicine

Walter Robb
Vice President & General Manager
Medical Systems Operations
General Electric
Milwaukee, WI

Frederick C. Robbins
University Professor
Department of Epidemiology and Biostatistics
School of Medicine
Case Western Reserve University
Cleveland, OH

Frank E. Samuel, Jr.
President
Health Industry Manufacturers' Association

Rosemary Stevens
Professor
Department of History
and Sociology of Science
University of Pennsylvania

ACKNOWLEDGMENTS

OTA would like to thank the members of the advisory panel and the contractors who provided information for this assessment. In addition, OTA acknowledges the following individuals for their assistance in reviewing drafts or supplying information.

F. Alan Andersen
U.S. Food and Drug Administration

Benjamin J. Barnhart
U.S. Department of Energy

Fred Bergmann
National Institute of General Medical Sciences

David Botstein
Massachusetts Institute of Technology

Noël Bouck
Northwestern University School of Medicine

Michele Calos
Stanford University School of Medicine

L.L. Cavalli-Sforza
Stanford University School of Medicine

George Church
University of California at San Francisco

Mary Ann Danello
U.S. Food and Drug Administration

Amiram Daniel
Health Industry Manufacturers Association

Frederick J. DeSerres
National Institute of Environmental Health Sciences

Robert Dixon
U.S. Environmental Protection Agency

William Fitzsimmons
National Institute of General Medical Sciences

W. Gary Flamm
U.S. Food and Drug Administration

Joseph Fraumeni
National Cancer Institute

Helen Gee
National Institutes of Health

Margaret Gordon
National Institutes of Health

Ronald Hart
National Center for Toxicological Research

John Heddle
Ludwig Institute for Cancer Research (Canada)

Richard Hill
U.S. Environmental Protection Agency

Abraham Hsie
Oak Ridge National Laboratory

Seymour Jablon
National Research Council

Irene Jones
Lawrence Livermore National Laboratory

Moin Khoury
Johns Hopkins University

Norman Mansfield
National Institutes of Health

Karen Messing
University of Quebec at Montreal

Alexander Morley
Flinders Medical Center (Australia)

John J. Mulvihill
National Cancer Institute

DeLill Nasser
National Science Foundation

Neal S. Nelson
U.S. Environmental Protection Agency

Per Oftedal
University of Oslo

Doug Ohlendorf
E.I. du Pont de Nemours & Co.

Dilys Parry
National Cancer Institute

Claude Pickelsimer
Centers for Disease Control

William L. Russell
Oak Ridge National Laboratory

F. Raymond Salemme
E.I. du Pont de Nemours & Co.

Michael M. Simpson
Congressional Research Service

David A. Smith
U.S. Department of Energy

Gail Stetten
Johns Hopkins Hospital

J. Timothy Stout
Baylor College of Medicine

Gary Strauss
U.S. Environmental Protection Agency

Larry R. Valcovic
U.S. Environmental Protection Agency

Peter E. Voytek
U.S. Environmental Protection Agency

References

References

1. Albertini, R.J., "Studies With T-Lymphocytes: An Approach to Human Mutagenicity Monitoring," *Indicators of Genotoxic Exposure,* Banbury Report 13, B.A. Bridges, B.E. Butterworth, I.B. Weinstein (eds.) (Cold Spring Harbor, NY: Cold Spring Harbor Laboratory, 1982), pp. 393-410.
2. Albertini, R.J., "Somatic Gene Mutations In Vivo as Indicated by the 6-Thioguanine Resistant T-Lymphocytes in Human Blood," *Mutat. Res.* 150:411-422, 1985.
3. Albertini, R.J., Castle, K.L., and Borcherding, W.R., "T-Cell Cloning To Detect the Mutant 6-Thioguanine-Resistant Lymphocytes Present in Human Peripheral Blood," *Proc. Nat. Acad. Sci. (USA)* 79:6617-6621, 1982.
4. Albertini, R.J., O'Neill, J.P., Nicklas, J.A., et al., "Alterations of the *hprt* Gene in Human 6-Thioguanine Resistant T-Lymphocytes Arising In Vivo," *Nature* 316:369-371, 1985.
5. Altland, K., Kaempfer, M., Forssbohm, M., et al., "Monitoring for Changing Mutation Rates Using Blood Samples Submitted for PKU Screening," *Human Genetics, Part A: The Unfolding Genome* (New York: Alan R. Liss, Inc., 1982), pp. 277-287.
6. Awa, A.A., "Cytogenetic Study," *J. Radiat. Res. (Suppl.)* 16:75-81, 1975.
7. Awa, A.A., Sofuni, T., Honda, T., et al., "Preliminary Reanalysis of Radiation-Induced Chromosome Aberrations in Relation to Past and Newly Revised Dose Estimates for Hiroshima and Nagasaki A-Bomb Survivors," Radiation Effects Research Foundation Technical Report Series 8-83, 1983.
8. Awa, A.A., Honda, S., and Neriishi, et al., "An Interim Report of the Cytogenetic Study of the Offspring of Atomic Bomb Survivors in Hiroshima and Nagasaki," *Human Genetics, Part A: The Unfolding Genome* (New York: Alan R. Liss, Inc., 1982).
9. Bigbee, W.L., Langlois, R.G., Vanderlaan, M., et al., "Binding Specificities of Eight Monoclonal Antibodies to Human Glycophorin A—Studies With M^cM and MkEn(UK) Variant Human Erythrocytes and M- and MNv-type Chimpanzee Erythrocytes," *J. Immunol.* 133:3149-3155, 1984.
10. Bigbee, W.L., Vanderlaan, M., Fong, S.S.N., et al., "Monoclonal Antibodies Specific for the M- and N-Forms of Human Glycophorin A," *Molec. Immunol.* 20:1353-1362, 1983.
11. Boice, J.D., Jr., Day, N.E., and Andersen, A., et al., "Second Cancers Following Radiation Treatment for Cervical Cancer," *J. Nat. Cancer Inst.* 74(5):955-975, 1985.
12. Boué, A., Boué, J., and Gropp, A., "Cytogenetics of Pregnancy Wastage," *Adv. Hum. Genet.* 14:1-57, 1985.
13. Branscomb, E., "Detecting Genetic Injury in Body Cells: The Red Blood Cell as Sentinel," typescript, contract report prepared for the Office of Technology Assessment, U.S. Congress, Washington, DC, 1985.
14. Bridges, B.A., "An Approach to the Assessment of the Risk to Man From DNA Damaging Agents," *Arch. Toxicol.* Suppl.3:271-281, 1980.
15. Bridges, B.A., "Status and Prospects, in Indicators of Genotoxic Exposure," *Indicators of Genotoxic Exposure,* Banbury Report 13, B.A. Bridges, B.E. Butterworth, and I.B. Weinstein (eds.) (Cold Spring Harbor, NY: Cold Spring Harbor Laboratory, 1982), pp. 555-558.
16. Cariello, N., "Methods To Characterize Mutations in Human T-Lymphocytes," typescript, contract report prepared for the Office of Technology Assessment, U.S. Congress, Washington, DC, 1985.
17. Cavalli-Sforza, L.L., and Bodmer, W.F., *The Genetics of Human Populations* (San Francisco: W.H. Freeman & Co., 1971).
18. Chandley, A.D., "The Origin of Aneuploidy," *Human Genetics, Part B: Medical Aspects* (New York: Alan R. Liss, Inc., 1982), pp. 337-347.
19. Charles, D., and Pretsch, W., "A Mutation Affecting the Lactate Dehydrogenase Locus *LDH-1* in the Mouse. I: Genetical and Electrophoretic Characterization," *Biochem. Genet.* 19:301-307, 1981.
19a. Charles, D., and Pretsch, W., "Linear Dose-Response Relationship of Erythrocyte Enzyme Activity Mutations in Offspring of Ethylnitrosourea-Treated Mice," *Mutat. Res.*, in press.
20. Church, G.M., and Gilbert, W., "Genomic Sequencing," *Proc. Nat. Acad. Sci. (USA)* 81:1991-1995, 1984.
21. Costa, T., Scriver C.R., and Childs, B., "The Effect of Mendelian Disease on Human Health: A Measurement," *Amer. J. Med. Genet.* 21:231-242, 1985.
22. Crow, J.F., and Denniston, C., "Mutation in Human Populations," *Adv. Hum. Genet.* 14:59-123, 1985.

23. Czeizel, A., "Epidemiological Follow-Up Study on Mutagenic Effects in Self-Poisoning Persons," *Environmental Mutagens and Carcinogens*, T. Sugimura, S. Kondo, and H. Takebe (eds.) (New York and Tokyo: Alan R. Liss, Inc., and University of Tokyo Press, 1982).
24. Darby, S., Nakashima, E., and Kato, H., "A Parallel Analysis of Cancer Mortality Among Atomic Bomb Survivors and Patients With Ankylosing Spondylitis Given X-Ray Therapy," *J. Nat. Cancer Inst.* 75(1):1-19, 1985.
25. Delehanty, J., White, R.L., and Mendelsohn, M.L., "Approaches To Determining Mutation Rates in Human DNA," *Mutat. Res.* 167:215-232, 1986.
26. Dempsey, J.L., Seshadri, R.S., and Morley, A.A., "Incresased Mutation Frequency Following Treatment With Cancer Chemotherapy," *Cancer Res.* 45:2873-2877, 1985.
27. Dobson, R.L., and Felton, J.S., "Female Germ Cell Loss From Radiation and Chemical Exposures," *Am. J. Industrial Med.* 4:175-190, 1983.
28. Dobson, R.L., Kwan, T.C. and Straume, T., "Tritium Effects on Germ Cells and Fertility," *European Seminar on the Risks From Tritium Exposure*, G. Gerber and M. Myttenaere (eds.) (Brussels: Commission of the European Communities, 1984).
29. Dreyer, N.A., Kohn, H.I., Clapp, R., et al., *The Feasibility of Epidemiologic Investigations of the Health Effects of Low-Level Ionizing Radiation*, NUREG/CR-1728 (Washington, DC: U.S. Government Printing Office, 1980).
30. Eber, S.W., Dunnwald, M., and Heinemann, G., et al., "Prevalence of Partial Deficiency of Red Cell Triosephosphate Isomerase in Germany-A Study of 3,000 People," *Hum. Genet.* 67:336-339, 1984.
31. Ehling, U.H., "Methods To Estimate the Genetic Risk," *Mutations in Man*, Gunter Obe (ed.) (Berlin: Springer Verlag, 1984).
32. Emery, A.E.H., and Rimoin, D., "Nature and Incidence of Genetic Disease," *Principles and Practice of Medical Genetics,* A.E.H. Emery and D. Rimoin (eds.) (Edinburgh: Churchill Livingstone, 1983).
33. Emery, A.E.H., *An Introduction to Recombinant DNA* (Chichester: John Wiley, 1984).
34. Fischer, S.G., and Lerman L.S., "DNA Fragments Differing by Single Base-Pair Substitutions are Separated in Denaturing Gradient Gels: Correspondence With Melting Theory," *Proc. Nat. Acad. Sci. (USA)*: 80: 1579-83, 1983.
35. Generoso, W.M., Cain, K.T., Cacheiro, N.L.A., et al., "Response of Mouse Spermatogonial Stem Cells to X-Ray Induction of Heritable Reciprocal Translocations," *Mutat. Res.* 126:177-187, 1984.
36. Gunther, M., and Penrose, L.S., "The Genetics of Epiloia," *J. Genet.* 31:413-430, 1935.
37. Harris, H., *The Principles of Human Biochemical Genetics*, 4th ed. (Amsterdam: North-Holland, 1980).
38. Harris, H., Hopkinson, D.A., and Robson, E.B., "The Incidence of Rare Alleles Determining Electrophoretic Variants: Data on 43 Enzyme Loci in Man," *Ann. Hum. Genet.* 37:237-253, 1974.
39. Hassold, T.J., and Jacobs, P.A., "Trisomy in Man," *Ann. Rev. Genet.* 18:69-97, 1984.
40. Hayes, A., Costa, T., Scriver, C.R., et al., "The Effect of Mendelian Disease on Human Health, II: Response to Treatment," *Amer. J. Med. Genet.* 21:243-255, 1985.
41. Hitotsumachi, S., Carpenter, D.A., and Russell, W.L., "Dose-Repetition Increases the Mutagenic Effectiveness of N-ethyl-N-nitrosourea in Mouse Spermatogonia," *Proc. Nat. Acad. Sci. (USA)* 82:6619-6621, 1985. 7
42. Hook, E.B., "Prevalence of Chromosome Abnormalities During Human Gestation and Implications for Studies of Environmental Mutagens," *Lancet* 2: 169-172, 1981.
43. Hook, E.B., "Contribution of Chromosome Abnormalities to Human Morbidity and Mortality and Some Comments Upon Surveillance of Chromosome Mutation Rates," *Prog. Mut. Res.* 3:9-38, 1982.
44. Hook, E.B., and Cross, P.K., "Surveillance of Human Populations for Germinal Cytogenetic Mutations," *Environmental Mutagens and Carcinogens*, T. Sugimura, S. Kondo, and H. Takebe (eds.) (New York and Tokyo: Alan R. Liss, Inc., and University of Tokyo Press, 1982).
45. Hook, E.B., Schreinemachers, D.M., and Willey, A.M., et al., "Rates of Mutant Structural Chromosome Rearrangements in Human Fetuses: Data From Prenatal Cytogenetic Studies and Associations With Maternal Age and Parental Mutagen Exposure," *Amer. J. Hum. Genet.* 35:96-109, 1983.
46. International Commssion for Protection Against Environmental Mutagens and Carcinogens, Committee 4, Final Report, "Estimation of Genetic Risks and Increased Incidence of Genetic Disease Due to Environmental Mutagens," *Mutat. Res.* 115:225-291, 1983.
47. International Commission on Radiological Pro-

tection, *Recommendations of the International Commission on Radiological Protection*, ICRP Publication 26 (Oxford: Pergamon Press, 1977).

48. Jablon, S., National Research Council, personal communication, April 1986.
49. Jackson, E.R., Fowle, J.R., III, and Voytek, P., "Some Research Needs To Support Mutagenicity Risk Assessments From Whole Mammal Studies," *Utilization of Mammalian Specific Locus Studies in Hazard Evaluation and Estimation of Genetic Risk*, F.J. deSerres and W. Sheridan (eds.) (New York: Plenum Press, 1983), pp. 315-322.
50. Jensen, R.H., Bigbee, W., and Branscomb, E.W., "Somatic Mutations Detected by Immunofluorescence and Flow Cytometry," *Biological Dosimetry Cytometric Approaches to Mammalian Systems*, W.G. Eisert and M.L. Mendelsohn (eds.) (Berlin: Springer-Verlag, 1984), pp. 161-170.
51. Johnson, F.M., and Lewis, S.E., "Electrophoretically Detected Germinal Mutations Induced in the Mouse by Ethylnitrosourea," *Proc. Nat. Acad. Sci. (USA)* 78:3138-3141, 1981.
52. Johnson, F.M., and Lewis, S.E., "The Detection of ENU-Induced Mutants in Mice by Electrophoresis and the Problem of Evaluating the Mutation Rate Increase," *Utilization of Mammalian Specific Locus Studies in Hazard Evaluation and Estimation of Genetic Risk*, F.J. deSerres and W. Sheridan (eds.) (New York: Plenum Press, 1983), pp. 95-123.
53. Johnson, F.M., Roberts, G.T., Sharma, R.K., et al., "The Detection of Mutants in Mice by Electrophoresis: Results of a Model Induction Experiment With Procarbazine," *Genetics* 97:113-124, 1981.
54. Jones, I.M., Burkhart-Schultz, K., and Carrano, A.V., "A Study of the Frequency of Sister Chromatid Exchange and of Thioguanine Resistant Cells in Mouse Spleen Lymphocytes After *in vivo* Exposure to Ethylnitrosourea," *Mutat. Res.* 143:245-249, 1985.
55. Kato, H., and Schull W.J., "Cancer Mortality Among Atomic Bomb Survivors," *Life Span Study Report 9, Part 1*, Radiation Effects Research Foundation Technical Report 12-80, 1982.
56. Kato, H., Schull, W.J., and Neel, J.V., "Survival in Children of Parents Exposed to the Atomic Bomb, a Cohort-Type Study," *Amer. J. Hum. Genet.* 18:339, 1966.
57. Lerman, L.S., "A Proposal for the Detection of Mutations in a Large-Scale Sampling of the Human Genome Using Two-Dimensional Denaturing Gradient Separations," typescript, contract report prepared for the Office of Technology Assessment, U.S. Congress, Washington, D.C., 1985.
58. Livingston, G.K., "An Overview of Chromosomal, Micronucleus, Heritable Effects, Dominant Lethal, Heritable Translocation, and Specific Locus Tests," *Reproduction: The New Frontier in Occupational and Environmental Health Research*, J.E. Lockey, et al. (eds.) (New York: Alan R. Liss, Inc., 1984), pp. 417-427.
59. Lubs, H.A., and Ing, P.S., "Human Cytogenetic Nomenclature," *Principles and Practice of Medical Genetics*, A.E.H. Emery and D. Rimoin (eds.) (Edinburgh: Churchill Livingstone, 1983), p. 162.
60. Lyon, M.F., "Sensitivity of Various Germ Cell Stages to Environmental Mutagens," *Mutat. Res.* 87:323-345, 1981.
61. Lyon, M.F., "Problems in Extrapolation of Animal Data to Humans," *Utilization of Mammalian Specific Locus Studies in Hazard Evaluation and Estimation of Genetic Risk*, F.J. deSerres and W. Sheridan (eds.) (New York: Plenum Press, 1983), pp. 55-69.
62. Lyon, M.F., "Measuring Mutation in Man," *Nature* 318: 315-316, 1985.
63. Malling, H.V., and Valcovic, L.R., "Biochemical Specific Locus Mutation System in Mice," *Arch. Toxicol.* 38:45-51, 1977.
64. Malling, H.V., "Perspectives in Mutagenesis," *Environ. Mutag.* 3:103-108, 1981.
65. Matsunaga, E., "Perspectives in Mutation Epidemiology. 1: Incidence and Prevalence of Genetic Disease (Excluding Chromosomal Aberrations) in Human Populations," *Mutat. Res.* 99:95-128, 1982.
66. Matsunaga, E., "Perspectives in Mutation Epidemiology. 5: Modern Medical Practice Versus Environmental Mutagens: Their Possible Dysgenic Impact," *Mutat. Res.* 11:449-457, 1983.
67. Maxam, A.M., and Gilbert, W., "A New Method for Sequencing DNA," *Proc. Nat. Acad. Sci. (USA)* 74:560-564, 1977.
68. Mendelsohn, M., "Prospects for Cellular Mutational Assays in Human Populations," *Assessment of Risk From Low-Level Exposure to Radiation and Chemicals*, A.D. Woodhead, et al. (eds.) (New York: Plenum Press, 1985), pp. 415-427.
69. Mendelsohn, M., Lawrence Livermore National Laboratory, personal communication, May 6, 1986.
70. Messing, K., and Bradley, W.E.C., "In Vivo Mutant Frequency Rises Amoung Breast Cancer Pa-

tients After Exposure to High Doses of Gamma Radiation," *Mutat. Res.* 152:107-112, 1985.

71. Miller, J.R., "Perspectives in Mutation Epidemiology. 4: General Principles and Considerations," *Mutat. Res.* 114:425-447, 1983.
72. Miller, O.J., "Chromosomal Basis of Inheritance," *Principles and Practice of Medical Genetics*, A.E.H. Emery and D. Rimoin (eds.) (Edinburgh: Churchill Livingstone, 1983), pp. 49-64.
73. Mohrenweiser, H.W., "Feasibility Study: Detection of Chemically Induced Mutation by Assay of Metabolic Characteristics," *Substitute Chemical Program. II: Toxicological Methods and Genetics Effects Workshop* (Washington, DC: U.S. Environmental Protection Agency, 1975).
74. Mohrenweiser, H.W., "Biochemical Approaches To Monitoring Human Populations for Germinal Mutation Rates. II: Enzyme Deficiency Variants as a Component of the Estimated Genetic Risk," *Utilization of Mammalian Specific Locus Studies in Hazard Evaluation and Estimation of Genetic Risk*, F.J. deSerres and W. Sheridan (eds.) (New York: Plenum Press, 1983) pp. 55-69.
75. Mohrenweiser, H.W., "Enzyme-Deficiency Variants: Frequency and Potential Significance in Human Populations," *Isozymes: Current Topics in Biological and Medical Research* 10:51-68, 1983.
76. Mohrenweiser, H.W., unpublished data, cited in Neel, J.V., Satoh, C., Goriki, K., et al., "The Rate With Which Spontaneous Mutation Alters the Electrophoretic Mobility of Polypeptides," *Proc. Nat. Acad. Sci. (USA)* 83:389-393, 1986.
77. Mohrenweiser, H.W., "One-Dimensional Electrophoresis and Quantitative Enzyme Assays as Technologies To Detect Mutation Rates in Human Beings," typescript, contract report prepared for the Office of Technology Assessment, U.S. Congress, Washington, DC, 1985.
78. Mohrenweiser, H.W., and Neel, J.V., "A 'Disproportion' Between the Frequency of Rare Electromorphs and Enzyme Deficiency Variants in Amerindians," *Amer. J. Hum. Genet.* 36:655-662, 1984.
79. Morley, A.A., Trainor, K.J., Seshadri, R., and Ryall, R.G., "Measurement of In Vivo Mutations in Human Lymphocytes," *Nature* 302:155-156, 1983.
80. Mukai, T., and Cockerham, C.C., "Spontaneous Mutation Rates at Enzyme Loci in *Drosophila melanogaster*," *Proc. Nat. Acad. Sci. (USA)* 74:2514-2517, 1977.
81. Müller, H.J., "Artificial Transmutation of the Gene," *Science* 66:84-87, 1927.
82. Mulvihill, J.J., and Byrne, J., "Offspring of Long-Time Survivors of Childhood Cancer," *Childhood Cancer: Late Effects, Clinics in Oncology* 4(2):333-343, 1985.
83. Mulvihill, J.J., and Czeizel, A., "Perspectives in Mutation Epidemology. 6: A 1983 View of Sentinel Phenotypes," *Mutat. Res.* 123:345-61, 1983.
84. Myers, R.M., "Determination of the Mutation Rate in Humans by Denaturing Gradient Gel Electrophoresis," typescript, contract report prepared for the Office of Technology Assessment, U.S. Congress, Washington, DC, 1985.
85. Myers, R.M., "Determination of the Mutation Rate in Humans by Ribonuclease Cleavage at Mismatches in RNA:DNA Heteroduplexes," typescript, contract report prepared for the Office of Technology Assessment, U.S. Congress, Washington, DC, 1985.
86. Myers, R.M., Larin, Z., and Maniatis, T., "Detection of Single Base Substitutions by Ribonuclease Cleavage at Mismatches in RNA:DNA Duplexes," *Science* 230:1242-1246, 1985.
87. National Research Council, Committee on the Biological Effects of Ionizing Radiation, *The Effects on Populations of Exposure to Low Levels of Ionizing Radiation* (Washington, DC: National Academy Press, 1980).
88. National Research Council, *Identifying and Estimating the Genetic Impact of Chemical Mutagens* (Washington, DC: National Academy Press, 1983).
89. Neel, J.V., "The Detection of Increased Mutation Rates in Human Populations," *Persp. Biol. Med.* 14:522-537, 1971.
90. Neel, J.V., "Some Trends in the Study of Spontaneous and Induced Mutation in Man," *Hum. Genet.*, Proc. 5th Intl. Cong., Mexico (Amsterdam: Excerpta Medica, 1977), pp. 19-32.
91. Neel, J.V., "Frequency of Spontaneous and Induced 'Point' Mutations in Higher Eukaryotes," *J. Hered.* 74:2-15, 1983.
92. Neel, J.V., University of Michigan Medical School, personal communication, Aug. 21, 1985.
93. Neel, J.V., Mohrenweiser, H., and Hanash S., et al., "Biochemical Approaches To Monitoring Human Populations for Germinal Mutation Rates. I: Electrophoresis," *Utilization of Mammalian Specific Locus Studies in Hazard Evaluation and Estimation of Genetic Risk*, F.J. deSerres and W. Sheridan (eds.) (New York: Plenum Press, 1983), pp. 71-93.
94. Neel, J.V., Mohrenweiser, H.W., and Meisler, M.H., "Rate of Spontaneous Mutation at Human Loci Encoding Protein Structure," *Proc. Nat. Acad. Sci. (USA)* 77(10)6037-6041, 1980.

95. Neel, J.V., Rosenblum, B.B., Sing, C.F., et al., "Adapting Two-Dimensional Gel Electrophoresis to the Study of Human Germ-Line Mutation Rates," ch. 9, *Two-Dimensional Gel Electrophoresis of Proteins*, J.E. Celis and R. Bravo (eds.) (London: Academic Press, 1984), pp. 259-306.

96. Neel, J.V., Satoh, C., Goriki, K., et al., "The Rate With Which Spontaneous Mutation Alters the Electrophoretic Mobility of Polypeptides," *Proc. Nat. Acad. Sci. (USA)* 83:389-393, 1986.

97. Nicklas, J., "The Use of T-Lymphocytes for Mutagenicity Monitoring," typescript, contract report prepared for the Office of Technology Assessment, U.S. Congress, Washington, DC, 1985.

98. Olson, M.V., "Detection of New Germinal Mutations by Surveying Human DNA for the Gain or Loss of Restriction-Enzyme Cleavage Sites," typescript, contract report prepared for the Office of Technology Assessment, U.S. Congress, Washington, DC, 1985.

99. Otake, M., and Schull, W.J., "In Utero Exposure to A-Bomb Radiation and Mental Retardation: A Reassessment," *Brit. J. Radiol.* 57:409-414, 1984.

100. Phillips, J.A., and Kazazian, H.H., "Principles and Practice of Medical Genetics," *Haemoglobinopathies and Thalassaemias*, A.E.H. Emery and D.L. Rimoin (eds.) (Edinburgh: Churchill Livingstone, 1983), pp. 1019-1043.

101. Porter, I.H., "Control of Hereditary Disorders," *Ann. Rev. Public Health* 3:277-319, 1982.

102. President's Commission for the Study of Ethical Problems in Medicine and Biomedical Research and Behavioral Research, *Screening and Counseling for Genetic Conditions* (Washington, DC: U.S. Government Printing Office, 1983).

103. Racine, R.R., Langley, C.H., and Voelker, R.A., "Enzyme Mutants Induced by Low-Dose Rate Gamma-Irradiation in *Drosophila*: Frequency and Characterization," *Environ. Mutag.* 2:167-177, 1980.

104. Rosenblum, B.B., Neel, J.V., and Hanash, S.M., "Two-Dimensional Electrophoresis of Plasma Polypeptides Reveals 'High' Heterozygosity Indices," *Proc. Nat. Acad. Sci. (USA)* 80:5002-5006, 1983.

105. Rosenblum, B.B., Neel, J.V., Hanash, S.M., et al., "Identification of Genetic Variants in Erythrocyte Lysate by Two-Dimensional Gel Electrophoresis," *Amer. J. Hum. Genet.* 36:601-612, 1984.

106. Russell, L.B., Aaron, C.S., de Serres, F., et al., "Evaluation of Mutagenicity Assays for Purposes of Genetic Risk Assessment," *Mutat. Res.* 134:143-157, 1984.

107. Russell, L.B., Russell, W.L., Popp, R.A., et al., "Radiation Induced Mutations at Mouse Hemoglobin Loci," *Proc. Nat. Acad. Sci. (USA)* 73:2843-2846, 1976.

108. Russell, L.B., and Shelby, M.D., "Tests for Heritable Genetic Damage and for Evidence of Gonadal Exposure in Mammals," *Mutat. Res.*, 154:69-84, 1985.

109. Russell, W.L., "X-Ray-Induced Mutations in Mice," *Cold Spring Harbor Symposium in Quantitative Biology* XVII:327-335, 1951.

110. Russell, W.L., "Mutation Frequencies in Female Mice and the Estimation of Genetic Hazards of Radiation in Women," *Proc. Nat. Acad. Sci. (USA)* 74:3521-3527, 1977.

111. Russell, W.L., "Dose Response, Repair, and No-Effect Dose Levels in Mouse Germ-Cell Mutagenesis," *Problems of Threshold in Chemical Mutagenesis*, Proceedings of a Nissan Symposium on Dose-Response Relationship for Genetic Effects of Environmental Chemicals, Tokyo, May 7-9, 1984, pp. 43-50.

112. Russell, W.L., Hunsicker, P.R., Carpenter, D.A., et al., "Effect of Dose Fractionation on the Ethylnitrosourea Induction Specific-Locus Mutations in Mouse Spermatogonia," *Proc. Nat. Acad. Sci. (USA)* 79:3592-3593, 1982.

113. Russell, W.L., and Kelly, E.M., "Mutation Frequencies in Male Mice and the Estimation of Genetic Hazards of Radiation in Men," *Proc. Nat. Acad. Sci. (USA)* 79:542-544, 1982.

114. Russell, W.L., and Kelly, E.M., "Specific-Locus Mutation Frequencies in Mouse Stem-Cell Spermatogonia at Very Low Radiation Dose Rates," *Proc. Nat. Acad. Sci. (USA)* 79:539-541, 1982.

115. Russell, W.L., Russell, L.B., and Kelly, E.M., "Radiation Dose Rate and Mutation Frequency," *Science* 128:1546-1550, 1958.

116. Sachs, E.S., Van Hemel, J.O., and Galjaard, H., et al., "First Trimester Chromosomal Analysis of Complex Structural Rearrangements With RHA Banding on Chorionic Villi" (letter), *Lancet* 2:1426, 1983.

117. Sanderson, B.J.S., Dempsey, J.L., and Morley, A.A., "Mutations in Human Lymphocytes: Effect of X- and UV-Irradiation," *Mutat. Res.* 140:223-227, 1984.

118. Sanger, F., Nicklen, S., and Coulson, A.R., "DNA Sequencing With Chain-Terminating Inhibitors," *Proc. Nat. Acad. Sci. (USA)* 74:5463-5467, 1977.

119. Sankaranarayanan, K., "The Role of Non-

Disjunction in Aneuploidy in Man: An Overview," *Mutat. Res.* 61:1-28, 1979.

120. Sankaranarayanan, K., *Genetic Effects of Ionizing Radiation in Multicellular Eukaryotes and the Assessment of Genetic Radiation Hazards in Man* (Amsterdam: Elsevier Biomedical Press, 1982).
121. Satoh, C., Awa, A.A., and Neel, J.V., et al., "Genetic Effects of Atomic Bombs," *Human Genetics, Part A: The Unfolding Genome* (New York: Alan R. Liss, Inc., 1982), pp. 267-276.
122. Satoh, C., Neel, J.V., and Yamashita, A., et al., "The Frequency Among Japanese of Heterozygotes for Deficiency Variants of 11 Enzymes," *Amer. J. Hum. Genet.* 35:656-674, 1983.
123. Schull, W.J., Neel, J.V., and Hashizume, A., "Further Observations on Sex Ratio Among Infants Born to Survivors of the Atomic Bombs," *Amer. J. Hum. Genet.* 18:328, 1966.
124. Schull, W.J., Otake, M., and Neel, J.V., "Genetic Effects of the Atomic Bombs: A Reappraisal," *Science* 213:1220-1227, 1981.
125. Schull, W.J., Otake, M.,and Neel, J.V., "Hiroshima and Nagasaki: A Reassessment of the Mutagenic Effect of Exposure to Ionizing Radiation," *Population and Biological Aspects of Human Mutation*, E.B. Hook and I.H. Porter (eds.) (New York: Academic Press, 1981).
126. Smith, C.L., "Purification, Specific Fragmentation, and Separation of Large Molecular Weight DNA Including Intact Chromosomes," typescript, contract report prepared for the Office of Technology Assessment, U.S. Congress, Washington, DC, 1985.
127. Smith, C.L., and Cantor, C.R., "Pulsed-Field Gel Electrophoresis of Large DNA Molecules," *Nature* 319:701-702, 1986.
128. Smith, H.O., "Nucleotide Sequence Specificity of Restriction Endonucleases," *Science* 205:455-462, 1979.
129. Sobels, F.H., "Evaluating the Mutagenic Potential of Chemicals—The Minimal Battery and Extrapolation Problems," *Arch. Toxicol.* 46:21-30, 1980.
130. Sobels, F.H., "The Parallelogram: An Indirect Approach for the Assessment of Genetic Risks From Chemical Mutagens," *Prog. Mut. Res.* 3:323-327, 1982.
131. Stamatoyannopoulos, G., and Nute, P.E., "De Novo Mutations Producing Unstable Hbs or HbsM. II: Direct Estimates of Minimum Nucleotide Mutation Rates in Man," *Hum. Genet.* 60:181-88, 1982.
132. Stamatoyannopoulos, G., and Nute, P., "Detection of Somatic Mutants of Hemoglobin," *Utilization of Mammalian Specific Locus Studies in Hazard Evaluation and Estimation of Genetic Risk,* F.J. de Serres and W. Sheridan (eds.) (New York: Plenum Press, 1983).
133. Stamatoyannopoulos, G., Nute, P., Lindsley, D., et al., "Somatic-Cell Mutation Monitoring System Based on Human Hemoglobin Mutants," *Single-Cell Mutation Monitoring Systems—Methodologies and Applications*, A.A. Ansari and F.J. deSerres (eds.) (New York: Plenum Press, 1984), pp. 1-35.
134. Stamatoyannopoulos, G., Nute, P.E., and Miller, M., "De Novo Mutations Producing Unstable Hbs or Hbs M. I: Establishment of a Depository and Use of Data To Test for an Association of De Novo Mutation With Advanced Parental Age," *Hum. Genet.* 58:396-404, 1981.
135. Stein, Z., Kline, J., and Susser, E., et al., "Maternal Age and Spontaneous Abortion," *Embryonic and Fetal Death*, I.H. Porter and E.B. Hook (eds.) (New York: Academic Press, 1980).
136. Stein, Z., Susser, M., and Warburton, D., et al., "Spontaneous Abortion as a Screening Device," *Amer. J. Epidem.* 102:275-290, 1975.
137. Strauss, G.H.S., "Direct Mutagenicity Testing: The Development of a Clonal Assay To Detect and Quantitate Mutant Lymphocytes Arising in Vivo," *Indicators of Genotxic Exposure,* Banbury Report 13, B.A. Bridges, B.E. Butterworth, and I.B. Weinstein (eds.) (Cold Spring Harbor, NY: Cold Spring Laboratories, 1982), pp. 423-439.
138. Streisinger, G., "Extrapolations From Species to Species and From Various Cell Types in Assessing Risks From Chemical Mutagens," *Mutat. Res.* 114:93-105, 1983.
139. Thilly, W.G., "Potential Use of Gradient Denaturing Gel Electrophoresis in Obtaining Mutational Spectra From Human Cells," *Carcinogenesis: The Role of Chemicals and Radiation in the Etiology of Cancer,* vol. 10, E. Huberman (ed.) (New York: Raven Press, 1985), pp. 511-528.
140. Tobari, Y.N., and Kojima, K., "A Study of Spontaneous Mutation Rates at Ten Loci Detectable by Starch Gel Electrophoresis in *Drosophila melanogaster,*" *Genetics* 70:397-403, 1972.
141. Trainor, K.J., Wigmore, D.J., Chrysostomou, A., et al., "Mutation Frequency in Human Lymphocytes Increases With Age," *Mech. Age. Develop.* 27:83-86, 1984.
142. Tucker, M.A., Meadows, A.T., Boice, J.D., Jr., "Cancer Risk Following Treatment of Childhood Cancer," *Radiation Carcinogenesis: Epidemiology and Biological Significance*, J.D. Boice, Jr., and J.F. Fraumeni, Jr. (eds.) (New York: Raven Press, 1984).

143. Turner, D.R., Morley, A.A., Haliandros, M., et al., "In Vivo Somatic Mutations in Human Lymphocytes Are Frequently Due to Major Gene Alterations," *Nature* 315:343-345, 1985.
144. UNSCEAR Report, *Sources and Effects of Ionizing Radiation*, Report of the United Nations Scientific Committee on the Effects of Atomic Radiation, United Nations, New York, 1982.
145. U.S. Congress, Office of Technology Assessment, *Assessment of Technologies for Determining Cancer Risks From the Environment*, OTA-H-138 (Washington, DC: U.S. Government Printing Office, June 1981).
146. U.S. Congress, Office of Technology Assessment, *Impacts of Applied Genetics: Micro-Organisms, Plants, and Animals*, OTA-HR-132 (Washington, DC: U.S. Government Printing Office, April 1981).
147. U.S. Congress, Office of Technology Assessment, *Commercial Biotechnology: An International Analysis*, OTA-BA-218 (Washington, DC: U.S. Government Printing Office, January 1984).
148. U.S. Department of Health and Human Services, Public Health Service, Food and Drug Administration, *The Selection of Patients for X-Ray Examination*, HHS Publication (FDA) 80-8104 (Washington, DC: U.S. Government Printing Office, January 1980).
149. U.S. Department of Health and Human Services, National Institutes of Health, National Cancer Institute, *Multiple Primary Cancers in Connecticut and Denmark*, National Cancer Insititute Monograph 68, NIH Publication 85-2714, J.C. Boice, et. al. (eds.) (Bethesda, MD: 1985.)
150. U.S. Department of Health and Human Services, National Institutes of Health, *Genetic Sequence Data Bank* (Cambridge, MA: Bolt, Beranek & Newman, Inc., 1985).
151. U.S. Department of Labor, Occupational Safety and Health Administration, "Occupational Exposure to Ethylene Dibromide: Notice of Proposed Rulemaking," *Federal Register* 48(196): 45956-46003, Oct. 7, 1983.
152. U.S. Department of Labor, Occupational Safety and Health Administration, "Occupational Exposure to Ethylene Oxide: Final Standard," *Federal Register* 49(122):25734-25809, June 22, 1984.
153. U.S. Environmental Protection Agency, "Federal Radiation Protection Guidance for Occupational Exposure," *Federal Register* 46(15), Jan. 23, 1981.
154. U.S. Environmental Protection Agency, "Proposed Guidelines for Mutagenicity Risk Assessment," *Federal Register* 9(227):46314-46321, Nov. 23, 1984.
155. U.S. Environmental Protection Agency, *Workshop Proceedings: Approaches for Improving the Assessment of Human Genetic Risk-Human Biomonitoring* (Washington, DC: 1984).
156. U.S. Environmental Protection Agency, Office of Radiation Programs, *Radionuclides: Background Information Document for Final Rules*, EPA 520/1-84-022-1 (Washington, DC: October 1984).
157. Van der Ploeg, L.H.T., "Determining Mutation Frequencies in Humans: The Use of Pulsed Field Gel-Electrophoresis for the Analysis of Chromosomes," typescript, contract report prepared for the Office of Technology Assessment, U.S. Congress, Washington, DC, 1985.
158. Van der Ploeg, L.H.T., Smits, M., Ponnudurai, T., et al., "Chromosome-Sized DNA Molecules of *Plasmodium Falciparum*," *Science* 229:658-661, 1985.
159. Van Dyke, D.L., Weiss, L., and Roberson, J.R., et al., "The Frequency and Mutation Rate of Balanced Autosomal Rearrangements in Man Estimated From Prenatal Genetic Studies for Advanced Maternal Age," *Amer. J. Hum. Genet.* 35:301-308, 1983.
160. Vijayalaxmi, and Evans, H.J., "Measurement of Spontaneous and X-Irradiation-Induced 6-Thioguanine-Resistant Human Blood Lymphocytes Using a T-Cell Cloning Technique," *Mutat. Res.* 125:87-94, 1984.
161. Voelker, R.A., Schaffer, H.E., and Mukai, T., "Spontaneous Allozyme Mutations in *Drosophila*: Rate of Occurrence and Nature on the Mutants," *Genetics* 94:961-968, 1980.
162. Vogel, F., "Mutation in Man," *Principles and Practice of Medical Genetics*, A.E.H. Emery and D. Rimoin (eds.) (Edinburgh: Churchill Livingstone, 1983), pp. 26-48.
163. Vogel, F., "Gene or Point Mutations," *Mutations in Man*, G. Obe (ed.) (Berlin: Springer-Verlag, 1984), pp. 101-127.
164. Vogel, F., and Altland, K., "Utilization of Material From PKU-Screening Programs for Mutation Screening," *Prog. Mut. Res.* 3:143-52, 1982.
165. Vogel, F., and Motulsky, A.G., *Human Genetics: Problems and Approaches* (Berlin: Springer-Verlag, 1982).
166. Vogel, F., and Rathenberg, R., "Spontaneous Mutation in Man," *Adv. Hum. Genet.* 5:223-318, 1975.
167. Voytek, P., "Approaches for Regulating Mutagenic Agents," *Reproduction: The New Frontier in Occupational and Environmental Health Research* (New York: Alan R. Liss, Inc., 1984), pp. 429-438.
168. Voytek, P., Reproductive Effects Assessment

Group, U.S. Environmental Protection Agency, personal communication, Aug. 7, 1985.

169. Warburton, D., Stein, Z., and Kline, J., et al., "Chromosome Abnormalities in Spontaneous Abortion: Data From the New York City Study," *Embryonic and Fetal Death*, I.H. Porter and E.B. Hook (eds.) (New York: Academic Press, 1980), pp. 261-287.
170. Waters, M., Allen, J., Doerr, C., et al., "Application of the Parallelogram Approach To Predict Genotoxic Effects in Humans," typescript, 1985.
171. Weatherall, D.J., *The New Genetics and Clinical Practice* (London: Nuffield Provincial Hospitals Trust, 1982).
172. Yang, T.P., Patel, P.I., Chinault, A.C., et al., "Molecular Evidence for New Mutation at the *hprt* Locus in Lesch-Nyhan Patients," *Nature* 310:412-414, 1984.

Index

Index

O